Lecture Notes in Mathematics   1711

Editors:
A. Dold, Heidelberg
F. Takens, Groningen
B. Teissier, Paris

Springer
Berlin
Heidelberg
New York
Barcelona
Hong Kong
London
Milan
Paris
Singapore
Tokyo

Werner Ricker

Operator Algebras Generated by Commuting Projections: A Vector Measure Approach

 Springer

Author

Werner Ricker
School of Mathematics
University of New South Wales
Sydney, NSW, 2052
Australia
e-mail: werner@maths.unsw.edu.au

Cataloging-in-Publication Data applied for

Die Deutsche Bibliothek - CIP-Einheitsaufnahme

Ricker, Werner:
Operator algebras generated by commuting projections: a vector
measure approach / Werner Ricker. - Berlin ; Heidelberg ; New York
; Barcelona ; Hong Kong ; London ; Milan ; Paris ; Singapore ;
Tokyo : Springer, 1999
  (Lecture notes in mathematics ; 1711)
  ISBN 3-540-66461-0

Mathematics Subject Classification (1991): 28B05, 06E15, 47B40, 47D30

ISSN 0075-8434
ISBN 3-540-66461-0 Springer-Verlag Berlin Heidelberg New York

Typesetting: Camera-ready T_EX output by the author
SPIN: 10700204   41/3143-543210 - Printed on acid-free paper

For Margit, Simon and Sandra; without their constant encouragement and support these notes would never have eventuated.

PREFACE

In the summer semester of 1997 Professor Jim Cooper invited me to give an advanced set of lectures at the Honours/Masters level in the Mathematisches Institut of the Johannes Kepler Universität in Linz, Austria. He left the choice of topic up to me; his only request was that the topic should be of an *interdisciplinary nature* since the students already had a good background in such *individual* courses as algebra, linear algebra, real and complex analysis, functional analysis and measure theory, etc.. The content of this book is essentially an expanded version of the lectures given in Linz. The material was chosen in an attempt to illustrate to advanced students that it is indeed possible to present lecture courses within their mathematical reach which form a connecting bridge between many of the specialized courses that they have already had and such that it "all comes together."

In addition to being able to absorb a body of mathematical knowledge (hopefully developed in a systematic and coherent way), students at this level should also become accustomed to the methodology of mathematical research. They should be able to go to libraries and consult research books and articles, extract from these the relevant information, do some independent thinking, come to the realization that not all problems have instant solutions, etc.. Accordingly, there are many references to the mathematical literature (which the reader is expected to follow up), both in the text and in the various exercises. The exercises are a mixture of fairly routine ones (indicated by [*]) to somewhat more challenging ones and form an integral part of the notes. This book is surely not a pure research book on the topic; for this we refer to the excellent monographs [13] and [15], for example. It is more of a hybrid and, for this reason, definitions and statements of results are carefully formulated and referenced, examples are included to illustrate various points, and many of the proofs are quite detailed as they are designed for the working student and future researcher, and not (necessarily) current experts. At the same time, several of the chapters contain a significant amount of material which may also interest current researchers in the area. Moreover, any reader who achieves a firm grasp of the material is well placed to begin serious research in the general area of operator theory alluded to in these notes, especially in some of the more recent directions. I have here in mind two general areas. Firstly, there is the extension of the theory to the setting of non-normable spaces, where genuine new phenomena arise which are not present in the Banach space setting. Even though much has already been achieved in this direction in the past 20-30 years (see the works of P.G. Dodds, C.K. Fong, J. Junggeburth, S. Okada, B. de Pagter, W.J. Ricker, A. Shuchat, W.V. Smith in the Appendix, and of C.

Ionescu Tulcea, F.Y. Maeda, H.H. Schaefer, B.J. Walsh listed in the bibliography of [15]), there are still several important problems which remain unresolved. The other area is in the direction of harmonic analysis and differential operators in Euclidean L^p-spaces, which often generate families of commuting projections based on algebras or δ-rings of sets rather than σ-algebras of sets; see the works of E. Albrecht, G.B. Folland, G. Gaudry, G.E. Huige, J. Locker, G. Mockenhaupt, M.A. Shobov, W.V. Smith, H.J. Sussmann, I.P. Syroid in the Appendix, and of V.E. Ljance, V.A. Marčenko, M.A. Naimark, B.S. Pavlov, J.T. Schwartz listed in the bibliography of [15]. Such families of projections are typically not uniformly bounded and so will not lead to a Borel functional calculus of the type usually associated with a spectral operator. New techniques will be needed to analyze the operator algebras that such families of projections generate.

Many of the results presented are classical so I have not attempted to record the source of every item. References are not always to the original source, but often to more recent works where further references can be found. The absence of a reference does not necessarily imply originality on my part.

The reader is assumed to have a basic grasp of standard undergraduate courses in algebra, linear algebra, set theory (manipulation), topology, functional analysis, measure theory and integration. Since not all of the readers will have a common base of knowledge in this regard, and for the reason of self-containment, some of these basic notions and facts are included. This is especially true of those having a direct bearing on the subject matter. More specialized material (eg. measure algebras, vector measures and integration, Stone spaces, aspects of operator algebras, functional calculi, etc.) is developed along the way, but only to the extent required for these notes.

In an Appendix at the end of the text I have attempted to form an extensive bibliography of research articles in the general area of spectral operators and Boolean algebras of projections which have appeared *since* 1979. For articles prior to 1979 we refer the interested reader to the excellent bibliographies in [13] and [15]. Some relevant papers prior to 1979 have also been included, provided they do not occur in [13] or [15]. The reason for this Appendix is two-fold. First, it is always useful for any student and/or researcher to have access to such extensive and up-to-date bibliographies. Second, and perhaps more important, I wish to illustrate to students and future researchers that this is an *active* area of modern research. This can be seen not only from the number of articles and their diversity, but also from the number of mathematicians who have contributed to the area.

Special thanks go to my colleagues and friends J.B. Cooper, K. Kiener, E. Matoušková and C. Stegall from Linz. Their encouragement, attention, assistance and above all, their patience, were remarkable. To all of my many colleagues and friends over the past years who have, at various stages, listened to my thoughts and ramblings on this topic (both directly and indirectly) and who have made helpful suggestions (both positive and negative), I especially wish to thank E. Albrecht, R.G. Bartle, I.D. Berg, E. Berkson, P.G. Dodds, I. Doust, D.H. Fremlin, T.A. Gillespie, D. Hadwin, B.R.F. Jefferies, I. Kluvánek, H.P. Lotz, A. McIntosh,

S. Okada, M. Orhon, B. de Pagter, F. Räbiger, P. Ressel, H.H. Schaefer, J.J. Uhr Jr., and A.I. Veksler. Finally I wish to thank Mrs J. Kos and Ms V. Pratto, both for their excellent typing and for their unlimited tolerance and understanding, and Dr P. Blennerhassett for his expert assistance in several of the finer points of LaTeX.

Sydney; June, 1999

Contents

Contents

INTRODUCTION

One of the fundamental facts learnt in linear algebra courses is a basic structural result referred to as the Jordan decomposition theorem. Namely, in a finite dimensional vector space X every linear map $T : X \to X$ can be decomposed as $T = S + N$, where S is a *diagonalizable operator* (i.e. with respect to a suitable basis of X it is similar to a diagonal operator) and N is a *nilpotent operator* (i.e. the spectrum $\sigma(N)$, of N, consists just of $\{0\}$ or, equivalently, $N^k = 0$ for some non-negative integer k) satisfying $SN = NS$. The operator S is called the *scalar part* of T and N is called the *radical part* of T. In particular, S has a representation of the form

$$(1) \qquad S = \sum_{j=1}^{r} \lambda_j E_j \,,$$

where $\sigma(S) = \sigma(T) = \{\lambda_j\}_{j=1}^r$ consists of the distinct eigenvalues of S and $\{E_j\}_{j=1}^r$ is a family of non-zero projections (i.e. $E_j^2 = E_j$) with $\sum_{j=1}^r E_j = I$ (the identity operator on X) and satisfying $E_j E_k = 0 = E_k E_j$ whenever $j \neq k$. So, the study of such scalar operators S reduces to a study of the family of much simpler operators $\lambda_j E_j$, for $1 \le j \le r$. In fact, if $X_j = E_j X$ is the range of E_j, then the family of vector subspaces $\{X_j\}_{j=1}^r$ has the properties that $X_j \cap X_k = \{0\}$ for $j \neq k$, that $X_1 \oplus \ldots \oplus X_r = X$ and that $SX_j \subseteq X_j$, for $1 \le j \le r$. In particular, S restricted to X_j (which is the same as $\lambda_j E_j$ restricted to X_j) acts like $\lambda_j I_j$ in X_j, where $I_j : X_j \longrightarrow X_j$ is the identity operator. For an elegant and succinct account of this topic in terms of linear operators (rather than the usual matrix approach) we refer to [14; Chapter VII, Sections 1 & 2].

What happens if X is infinite dimensional and the linear operator S is continuous? Consider first the case when X is a Hilbert space. If S is *compact* and *normal* (or *selfadjoint*), then the classical spectral theorem of D. Hilbert asserts that $\sigma(S) = \{0\} \cup \{\lambda_j\}_{j=1}^\infty$ is a countable set in $\mathbb{C}$ (or $\mathbb{R}$) with $\lim_{n \to \infty} \lambda_n = 0$ (in the case when $\sigma(S)$ is infinite) and S has a representation of the form (compare with (1))

$$(2) \qquad S = \sum_{j=1}^{\infty} \lambda_j E_j \,,$$

where the commuting family of non-zero, selfadjoint projections $\{E_j\}_{j=1}^\infty$ is pairwise disjoint and satisfies $\sum_{j=0}^\infty E_j = I$; here E_0 is the orthogonal projection of X onto $\{x \in X : Sx = 0\}$.

The series (2) and the series $\sum_{j=0}^{\infty} E_j = I$ both converge in the *strong operator topology*. Removing the compactness requirement on S has the effect that $\sigma(S)$ may no longer be discrete. Indeed, $\sigma(S)$ can then be any compact subset of $\mathbb{C}$ (or $\mathbb{R}$ if S is selfadjoint). Moreover, to every Borel set $A \subseteq \mathbb{C}$ (the σ-algebra of all such sets is denoted by $Bo(\mathbb{C})$) there corresponds a selfadjoint projection $E(A)$ such that $E(\emptyset) = 0$, $E(\mathbb{C}) = I$ and the projections in the range $E(Bo(\mathbb{C}))$ of E satisfy

(3) $$E(A)E(B) = E(A \cap B) = E(B)E(A), \qquad A, B \in Bo(\mathbb{C}),$$

and

(4) $$\sum_{n=1}^{\infty} E(A_n) = E(\cup_{n=1}^{\infty} A_n),$$

whenever $\{A_n\}_{n=1}^{\infty} \subseteq Bo(\mathbb{C})$ are pairwise disjoint sets. Of course, the series (4) again converges in the strong operator topology. The condition (4) says that $A \mapsto E(A)$ is a *projection-valued measure* on $Bo(\mathbb{C})$. What is the analogue of (2)? Adopting the naive approach that integrals usually replace sums (in the "limit") suggests that

(5) $$S = \int_{\mathbb{C}} \lambda \, dE(\lambda) = \int_{\sigma(S)} \lambda \, dE(\lambda),$$

where the operator-valued integral (5) needs to be suitably defined. This turns out to indeed be the case and (5) is a formulation of the classical *spectral theorem* for arbitrary normal (or selfadjoint) operators.

The important features from the abstract point of view are that S is synthesized from a certain family of projections $\{E(A) : A \in Bo(\mathbb{C})\}$ via an integral formula of the type

$$S = \int_{\mathbb{C}} f(\lambda) \, dE(\lambda),$$

where $f(\lambda) = \lambda$, for $\lambda \in \mathbb{C}$. Moreover, the multiplicative property (3) of E implies that

$$S^n = \int_{\mathbb{C}} \lambda^n \, dE(\lambda) = \int_{\mathbb{C}} f(\lambda)^n \, dE(\lambda), \qquad n = 0, 1, 2, \ldots,$$

and more generally, that

$$g(S) := \int_{\mathbb{C}} g(\lambda) \, dE(\lambda)$$

for any Borel measurable function $g : \mathbb{C} \longrightarrow \mathbb{C}$ which is bounded on $\sigma(S)$. Actually, $\sigma(S)$ turns out to be the *support* of the measure E. So, all reasonable operators which are "functions of S", that is, operators of the form $g(S)$ for suitable g, are built up from the projections $\{E(A) : A \in Bo(\mathbb{C})\}$.

If we wish to stay within the realm of normal operators, then it is necessary to require $\{E(A) : E \in Bo(\mathbb{C})\}$ to be a selfadjoint family. However, the properties (3) and (4) are independent of selfadjointness and so it is undesirable to require this condition from the outset. Moreover, removing this property is no great restriction. Indeed, the well known Mackey-Wermer theorem asserts that if the integral in (5) exists for an arbitrary projection-valued measure E (with respect to the strong operator topology), then there exists a selfadjoint isomorphism $W : X \to X$ such that the family of commuting projections $\{WE(A)W^{-1} : A \in Bo(\mathbb{C})\}$ consists entirely of selfadjoint projections. So, the infinite dimensional Hilbert space analogue of a scalar operator (still called a scalar operator) is any continuous linear operator S which is similar to a normal operator, in which case it has an integral representation of the form (5) for some projection-valued measure E defined on $Bo(\mathbb{C})$. The analogue of a nilpotent operator N is still one which satisfies $\sigma(N) = \{0\}$. However, in infinite dimensional spaces this becomes equivalent to $\lim_{n \to \infty} \|N^n\|^{1/n} = 0$, rather than to some power of N being 0; such operators are called *quasinilpotent*. So, a natural class of continuous linear operators in an infinite dimensional Hilbert space which corresponds to the familiar class of all linear operators in a finite dimensional space, consists of those operators T which have a decomposition

$$(6) \qquad T = S + N = \int_{\sigma(T)} \lambda \, dE(\lambda) + N,$$

where S is a scalar operator and N is a quasinilpotent operator satisfying $SN = NS$. In this formulation we see that even the Hilbert space structure of X is no longer crucial; the definitions of a scalar operator and quasinilpotent operator make perfectly good sense in a general Banach space X. In this setting, operators T of the form (6) are called *spectral operators*. This important class of operators, initiated by N. Dunford in the late 1940's and early 1950's, has undergone intense research ever since.

The aim of these notes is to concentrate on certain particular aspects of the theory of scalar operators, especially in the *Banach space setting*, where the results and methods differ significantly from those in the Hilbert space setting. As discussed above, the central notion is the family of projections $\mathcal{B} = \{E(A) : A \in Bo(\mathbb{C})\}$, the so called *resolution of the identity*, from which the scalar operator S is synthesized. However, to insist on indexing the projections in $\mathcal{B}$ by elements of $Bo(\mathbb{C})$ is, from the theoretical and practical viewpoint, both unnecessary and unduly restrictive. So, the basic concept throughout will be that of a family of commuting projections $\mathcal{B}$, assumed to form a *Boolean algebra* but otherwise not indexed in any particular way. Since we will be interested in those operators which can be "built up" from the elements of the Boolean algebra $\mathcal{B}$, it is natural to require the linear span of $\mathcal{B}$ to be an *algebra* (not just a vector space) and, since some limiting procedures will have to be involved (to pass from sums to integrals, for example), it will also be necessary to take the *closure* of this linear span with respect to some suitable topology. Moreover, to have any hope of identifying elements which arise as some sort of limit from expressions of the form $\sum_{j=1}^n \mu_j E_j$, where $\mu_j \in \mathbb{C}$ and $E_j E_k = 0 = E_k E_j$ if $j \neq k$, it is also a necessity to require $\sup\{\|E\| : E \in \mathcal{B}\}$ to be finite; this condition is automatic if $\mathcal{B}$ consists of selfadjoint projections in a Hilbert space, but not in general.

So, we arrive at the following setting: given is a Banach space X and a commutative, unital subalgebra $\mathcal{U}$ (of continuous linear operators on X) which is closed with respect to some topology and is generated by some Boolean algebra of projections $\mathcal{B}$ (assumed to be uniformly bounded). Our main purpose is to investigate, systematically and in detail, the theory of such operator algebras and to attempt to answer various natural questions. As a sample, we will consider the following problems.

(i)   Is it possible to give a concrete description of the elements of $\mathcal{U}$ in terms of those from $\mathcal{B}$? The answer will depend on various factors; the properties of the underlying Banach space X, the topology used in $\mathcal{U}$, and on certain properties of $\mathcal{B}$ itself. This question is the central theme of Chapter III, where the uniform operator topology is considered, and of Chapter V, where the strong and weak operator topologies are relevant.

(ii)   Are the elements of $\mathcal{U}$ all of the form $g(S)$ for suitable functions g and some scalar operator S? The important ingredients here turn out to be the "size" of $\mathcal{B}$ and certain properties of the Banach space X. One of the main results will be to show that the answer is affirmative if the Boolean algebra $\mathcal{B}$ is complete in a certain sense and if X is separable. This forms the core of Chapter VI and is a far reaching extension of the well known fact that every strongly closed Boolean algebra of selfadjoint projections in a separable Hilbert space is the resolution of the identity of some selfadjoint operator.

(iii)   Are there other descriptions of the elements of $\mathcal{U}$ with a more algebraic flavour? For instance, if X is a Hilbert space, then a classical result due to J. von Neumann provides a positive answer in terms of the bicommutant of $\mathcal{B}$ (provided that $\mathcal{B}$ consists of selfadjoint projections). Other descriptions are known in terms of the lattice of closed, $\mathcal{B}$-invariant subspaces of X. A detailed discussion of this topic is presented in Chapter VII.

Questions such as those above, and many more, were considered by N. Dunford and others. Several of the major results (but, certainly not all) concerning such operator algebras can be found in two penetrating papers by W.G. Bade [1,2]. These results, and others, are well documented in [13] and [15], for example. Anyone who spends time reading these monographs will realize immediately the beautiful combination of methods employed from a variety of areas within mathematics. From algebra we see the theory of partial orders, Boolean algebras and the representation results of M.H. Stone (as a sample), from functional analysis there is Banach algebra theory, functional calculi, Banach space geometry, weak and weak-star topologies, Alaoglu's theorem and so on, from measure theory we have the Riesz representation theorem, the Radon-Nikodym theorem, the Hahn decomposition theorem, operator-valued integrals and so on, from topology there occur various disconnected spaces, Urysohn's extension theorem, the Stone-Čech compactification, etc. etc.. So, there is no question that we are dealing with an "interdisciplinary topic".

In discussing commutative operator algebras which are uniform operator closed it is natural to employ Banach algebra techniques (as is the case in [15]). However, such methods are not always suitable to describe the *strongly* or *weakly* closed algebra generated by a Boolean algebra of projections. One of our main goals is to systematically employ the methods of *vector measures and integration theory* (developed in Chapter I to the extent needed for our purposes) to represent this algebra as an L^1-space of a *spectral measure*. Once this representation theory is available many of the results alluded to above are easy and natural consequences. In particular, our approach yields proofs of several of the well known theorems in the area which are quite different to the proofs given in [15].

That vector measure techniques can be employed at all relies on the fact that any Boolean algebra of projections B (with suitable completeness properties) can be realized as the range of a spectral measure defined on the *Baire or Borel sets* of the Stone space of B. This subtle interplay between Boolean algebras of projections and spectral measures, which plays a crucial and unifying role throughout these notes, is carefully developed in Chapter IV. To fully appreciate this subtle connection it is necessary to first consider general Boolean algebras (i.e. not necessarily consisting of projections on some Banach space) and their representation via the *closed-open* subsets of some totally disconnected, compact Hausdorff space. It turns out that the σ-algebra generated by these closed-open sets is precisely the family of *Baire sets*. Typically, the Baire sets form a *proper* sub-σ-algebra of the σ-algebra of all *Borel sets*. All of these features (and more) form the subject matter of Chapter II.

In conclusion, I wish to make it clear that the material presented here forms a personal choice of topics taken from a rather extensive area of research. I have not even attempted to touch on the theory of *spectral* operators, unbounded operators of scalar type, multiplicity theory, sums and products of commuting spectral operators, and so on. For this I refer the interested reader to [13], [15] and to the vast research literature on these topics which has appeared since the publication of [13] and [15], most of which is recorded in the Appendix.

Chapter I

Vector measures and Banach spaces

The first half of this chapter recalls some important notions and basic facts from classical (scalar-valued) measure theory and functional analysis. The second half of the chapter introduces vector measures (with values in a Banach space) and develops the theory of integration with respect to such measures, but only to the extent required in the sequel. Special emphasis is given to the usual convergence theorems and the L^1-space of a vector measure. The reader who is interested in more recent developments of such L^1-spaces should consult the works of G. Curbera [5, 6, 7]; these aspects of the theory will not be needed in these notes. Much of the basic theory of integration with respect to vector measures in *real* Banach spaces (and more general spaces) can be found in [27]. However, we wish to work in *complex* Banach spaces. Since the results for complex spaces do not always follow easily or directly from those for real spaces, we have decided to develop the theory for complex spaces directly. Many of these results can be found in [29] and others, such as the completeness of L^1, are new. Throughout this chapter, and the remainder of the text, the symbols $\mathbb{N}, \mathbb{R}$ and $\mathbb{C}$ will always denote the natural numbers $\{1, 2 \ldots\}$, the real numbers, and the complex numbers, respectively. So, let us begin.

Let Ω be a non-empty set. A family of subsets Σ of Ω is called an *algebra (of sets)* if

(i) $\Omega \in \Sigma$ and $\emptyset \in \Sigma$ (where $\emptyset$ denotes the empty set),

(ii) $E^c := \Omega \backslash E$ belongs to Σ whenever $E \in \Sigma$, and

(iii) $\cap_{j \in \mathcal{F}} E_j \in \Sigma$ for every finite collection $\{E_j : j \in \mathcal{F}\} \subseteq \Sigma$.

If Σ is an algebra of sets with the additional property that $\cap_{n=1}^{\infty} E_n \in \Sigma$ for every sequence $\{E_n\}_{n=1}^{\infty} \subseteq \Sigma$, then it is called a *$\sigma$-algebra*. In this case the pair (Ω, Σ) is called a *measurable space*.

Let (Ω, Σ) be a measurable space. A function $\nu : \Sigma \longrightarrow \mathbb{C}$ is called a *complex measure* if $\nu(\cup_{n=1}^{\infty} E_n) = \sum_{n=1}^{\infty} \nu(E_n)$ whenever $\{E_n\}_{n=1}^{\infty} \subseteq \Sigma$ is a sequence of pairwise disjoint sets, meaning that $E_n \cap E_m = \emptyset$ whenever $n \neq m$. We say that ν is *σ-additive*. In this case the triple (Ω, Σ, ν) is called simply a *measure space*. Since $\cup_{n=1}^{\infty} E_n = \cup_{n=1}^{\infty} E_{\pi(n)}$, for every bijection $\pi : \mathbb{N} \rightarrow \mathbb{N}$, it follows from the σ-additivity requirement that $\nu(\cup_{n=1}^{\infty} E_n) = \sum_{n=1}^{\infty} \nu(E_{\pi(n)})$. Accordingly, the series $\sum_{n=1}^{\infty} \nu(E_n)$ is necessarily *unconditionally convergent* in $\mathbb{C}$. Hence, it is actually *absolutely convergent*, meaning that $\sum_{n=1}^{\infty} |\nu(E_n)|$ is finite.

Whenever we write (Ω, Σ, ν) it is meant that Σ is a σ-algebra of subsets of a non-empty set Ω and $\nu : \Sigma \longrightarrow \mathbb{C}$ is a complex measure.

Example 1. (a) Let $\Omega = [0,1]$ and Σ denote the Borel sets in Ω, by definition the smallest σ-algebra containing all the open subsets of Ω. Define $\nu : \Sigma \longrightarrow \mathbb{C}$ by

$$\nu(E) := \sum_{n=1}^{\infty} \frac{i^n}{n(n+1)} \chi_E(\frac{1}{n}), \qquad E \in \Sigma.$$

Then ν is a complex measure. Here $\chi_E(t) := 1$ if $t \in E$ and $\chi_E(t) := 0$ if $t \notin E$.

(b) Let (Ω, Σ) be as in (a) and define $\nu : \Sigma \longrightarrow \mathbb{C}$ by

$$\nu(E) := \int_E \frac{1}{\sqrt{t}} \, dt, \qquad E \in \Sigma.$$

Then ν is a complex measure.

(c) Let $\Omega = [0, \infty)$ and Σ denote the Borel subsets of Ω. Then the usual Lebesgue measure $\nu : \Sigma \longrightarrow [0, \infty]$ is *not* a complex measure as it takes the "value" $\infty \notin \mathbb{C}$. ∎

Exercise 1.[∗] (a) Let $\Omega = \mathbb{N}$ be the natural numbers and Σ be the family of all subsets of Ω which are finite or have finite complement. Show that Σ is an *algebra (of sets)* . Is it a σ-algebra? Give reasons.

(b) Let $\Omega = [0,1]$ and Σ be the family of all subsets of Ω which are countable or have countable complement. Decide whether or not Σ is a σ-algebra. ∎

Exercise 2.   Let Σ be a σ-algebra of subsets of a set of $\Omega \neq \emptyset$ and let $\nu : \Sigma \longrightarrow \mathbb{C}$ be a complex, *finitely additive measure*, i.e. $\nu(\cup_{n=1}^k E_n) = \sum_{n=1}^k \nu(E_n)$ for all finite collections $\{E_n\}_{n=1}^k \subseteq \Sigma$ of pairwise disjoint sets. .

(a) Show that $\nu(A \backslash B) = \nu(A) - \nu(B)$ for all $A, B \in \Sigma$ with $B \subseteq A$. Here $A \backslash B := \{w \in A : w \notin B\}$.

(b) Show that ν is σ-additive if and only if $\lim_{n \to \infty} \nu(A_n) = 0$ whenever $\{A_n\}_{n=1}^{\infty} \subseteq \Sigma$ is a sequence of sets decreasing to $\emptyset$ (i.e. $A_1 \supseteq A_2 \supseteq A_3 \ldots$ and $\cap_{n=1}^{\infty} A_n = \emptyset$). ∎

Definition I.1   Let (Ω, Σ, ν) be a measure space . Define a function $|\nu| : \Sigma \longrightarrow [0, \infty)$ by

$$|\nu|(E) = \sup_{\pi} \sum_{A \in \pi} |\nu(A)|, \qquad E \in \Sigma,$$

where the supremum is taken over all finite partitions $\pi = \{F_1, \ldots, F_n\}$ of E in Σ, that is, $F_j \in \Sigma$, $F_j \cap F_k = \emptyset$ if $j \neq k$, $\cup_{j=1}^n F_j = E$ and $n \in \mathbb{N}$. The function $|\nu| : \Sigma \longrightarrow [0, \infty)$ is called the *variation measure* of ν.

Note that if we allow $|\nu|$ to take the "value" ∞, then $|\nu|$ is also defined for finitely additive measures $\nu : \Sigma \longrightarrow \mathbb{C}$, even if Σ is merely an algebra of sets. ∎

The following result summarizes some important features of complex measures and their variation; see [38; Chapter 6], [14; Chapter III], for example.

Theorem I.1.   *Let (Ω, Σ, ν) be a measure space , i.e. ν is σ-additive.*

(a) $|\nu|(\Omega) < \infty$.

(b) *The variation measure* $|\nu| : \Sigma \longrightarrow [0, \infty)$ *is a σ-additive, finite-valued measure.*

(c) $|\nu(E)| \leq |\nu|(E)$, *for all* $E \in \Sigma$.

(d) *If* $\mu : \Sigma \longrightarrow [0, \infty)$ *is any finitely additive measure satisfying* $|\nu(E)| \leq \mu(E)$, *for all* $E \in \Sigma$, *then* $|\nu|(E) \leq \mu(E)$ *for all* $E \in \Sigma$.

(e) $|\nu|(E) \leq |\nu|(F)$, *for all* $E, F \in \Sigma$ *with* $E \subseteq F$.

(f) $\sup\{|\nu(H)| : E \supseteq H \in \Sigma\} \leq |\nu|(E) \leq 4\sup\{|\nu(H)| : E \supseteq H \in \Sigma\}, \quad E \in \Sigma$.

The number $\|\nu\| := |\nu|(\Omega)$ is called the *total variation* of ν. In particular, the range $\nu(\Sigma) := \{\nu(E) : E \in \Sigma\}$ of any complex measure $\nu : \Sigma \longrightarrow \mathbb{C}$ is necessarily a bounded subset of $\mathbb{C}$.

For basic references to the theory of *topological spaces* we refer to [14], [24] or [37], for example. Suppose now that Ω is a *compact*, topological Hausdorff space. Let $C(\Omega)$ denote the set of all continuous, $\mathbb{C}$-valued functions on Ω. Given $f, g \in C(\Omega)$ and $\alpha \in \mathbb{C}$ we define functions $f + g$ and αf on Ω by $w \mapsto f(w) + g(w)$ and $w \mapsto \alpha f(w)$, for $w \in \Omega$, respectively. Then $f + g$ and αf are also elements of $C(\Omega)$ and so $C(\Omega)$ is a *vector space*. Define a function $\|\cdot\|_\infty : C(\Omega) \longrightarrow [0, \infty)$ by

$$\|f\|_\infty := \sup\{|f(w)| : w \in \Omega\}, \qquad f \in C(\Omega).$$

It is routine to check that

(i) $\|f\|_\infty = 0$ if and only if $f = 0$,

(ii) $\|\alpha f\|_\infty = |\alpha| \cdot \|f\|_\infty$, for all $\alpha \in \mathbb{C}$ and $f \in C(\Omega)$, and

(iii) $\|f + g\|_\infty \leq \|f\|_\infty + \|g\|_\infty$ for all $f, g \in C(\Omega)$.

Any function $\|\cdot\| : X \longrightarrow [0, \infty)$ defined on a vector space X which has the properties (i)–(iii) of $\|\cdot\|_\infty$ on $C(\Omega)$ listed above is called a *norm* on X. Recall that a sequence $\{x_n\}_{n=1}^\infty$ in a normed space $(X, \|\cdot\|)$ is a *Cauchy sequence* if for every $\varepsilon > 0$ there is a positive integer N_ε such that

$$\|x_n - x_m\| < \varepsilon, \quad \text{for all } m \geq N_\varepsilon \text{ and } n \geq N_\varepsilon.$$

If, for *every* Cauchy sequence $\{x_n\}_{n=1}^\infty$ in X there exists a vector $x \in X$ such that $\lim_{n\to\infty} x_n = x$, then we say that $(X, \|\cdot\|)$ is *complete* and call it a *Banach space*.

The space $(C(\Omega), \|\cdot\|_\infty)$ defined above is a well known example of a Banach space.

Example 2. (a) Every Hilbert space is a Banach space.

(b) Let c_0 denote the space of all sequences $\phi = (\phi_1, \phi_2, \dots)$ with complex entries and satisfying $\lim_{n\to\infty} \phi_n = 0$. Then c_0 is a vector space with respect to the co-ordinatewise operations $\psi + \phi := (\psi_1 + \phi_1, \phi_2 + \psi_2, \dots)$ and $\alpha\phi := (\alpha\phi_1, \alpha\phi_2, \dots)$ for each $\alpha \in \mathbb{C}$ and $\phi, \psi \in c_0$. A norm is defined in c_0 by

$$\|\phi\| := \sup\{|\phi_n| : n \in \mathbb{N}\}, \qquad \phi \in c_0,$$

with respect to which c_0 becomes a Banach space.

(c) Let c_{00} denote the collection of all elements $\phi \in c_0$ which have the property that there exists an integer $N_\phi > 0$ such that $\phi_n = 0$ for all $n \geq N_\phi$. Then c_{00} is a vector subspace of c_0 and so is a normed space with respect to the norm of c_0. The normed space $(c_{00}, \|\cdot\|)$ is *not* a Banach space as it fails to be complete.   ∎

Exercise 3. Verify that c_0 is complete, but c_{00} fails to be complete.   ∎

Let $(X, \|\cdot\|)$ be a Banach space. A map $\psi : X \longrightarrow \mathbb{C}$ which satisfies

$$\psi(\alpha x + \beta y) = \alpha \psi(x) + \beta \psi(y), \qquad \alpha, \beta \in \mathbb{C} \text{ and } x, y \in X,$$

is called a *linear functional* on X. If, in addition,

$$\|\psi\|_{X'} := \sup\{|\psi(x)| : x \in X, \|x\| \leq 1\}$$

is finite, then ψ is called *continuous* (or *bounded*). We also use the notation $\langle x, \psi \rangle := \psi(x)$, for $x \in X$. The space of all continuous linear functionals on X is called the (continuous) *dual space* of X and is denoted by X'.

Example 3. (a) Let $\Omega = [0,1]$ and $X = C(\Omega)$ with norm $\|\cdot\|_\infty$. Define $\psi : X \longrightarrow \mathbb{C}$ by

$$\langle f, \psi \rangle = f(\tfrac{1}{2}), \qquad f \in X.$$

Then $\psi \in X'$ and $\|\psi\| = 1$.

(b) Let $X = c_0$ with norm as defined in Example 2(b). Define $\psi : X \longrightarrow \mathbb{C}$ by

$$\langle \phi, \psi \rangle := \sum_{n=1}^{\infty} \frac{\phi(n)}{2^n}, \qquad \phi \in X.$$

Then $\psi \in X'$ and $\|\psi\| = 1$.   ∎

Let ψ_1, ψ_2 be continuous linear functionals on a Banach space X and $\alpha_1, \alpha_2 \in \mathbb{C}$. Then $\alpha_1 \psi_1 + \alpha_2 \psi_2 : X \longrightarrow \mathbb{C}$ defined by

$$\langle x, \alpha_1 \psi_1 + \alpha_2 \psi_2 \rangle := \alpha_1 \langle x, \psi_1 \rangle + \alpha_2 \langle x, \psi_2 \rangle, \qquad x \in X,$$

is a continuous linear functional on X as $\|\alpha_1 \psi_1 + \alpha_2 \psi_2\|_{X'} \leq |\alpha_1| \cdot \|\psi_1\|_{X'} + |\alpha_2| \cdot \|\psi_2\|_{X'}$. Hence, X' also becomes a vector space. It is a basic fact that X' is itself a Banach space for the norm $\|\cdot\|_{X'}$; see [14; Chapter II, §3], for example. Elements of X' are also traditionally denoted by x'.

Exercise 4. (a) Let $X = c_0$. Show if $\psi \in X'$, then there exists a sequence of complex numbers $\xi = (\xi_1, \xi_2, \dots)$ with $\sum_{n=1}^{\infty} |\xi_n| < \infty$ such that

$$\langle x, \psi \rangle = \sum_{n=1}^{\infty} x_n \xi_n, \qquad x = (x_1, x_2, \dots) \in X.$$

Moreover, $\|\psi\|_{X'} = \sum_{n=1}^{\infty} |\xi_n|$.

(b) Let $Y = \ell^1$ denote the space of all sequences of complex numbers $\xi = (\xi_1, \xi_2, \dots)$ for which $\|\xi\|_1 := \sum_{j=1}^{\infty} |\xi_j| < \infty$. Verify that $\|\cdot\|_1$ is a norm on Y and that $(Y, \|\cdot\|_1)$ is a Banach space. Show if $\Phi \in Y'$, then there exists a sequence of complex numbers $\rho = (\rho_1, \rho_2, \dots)$ with $\sup_n |\rho_n| < \infty$ such that

$$\langle \xi, \Phi \rangle = \sum_{n=1}^{\infty} \xi_n \rho_n, \qquad \xi = (\xi_1, \xi_2, \dots) \in Y.$$

Moreover, $\|\Phi\|_{Y'} = \sup_n |\rho_n|$.

(c) The space consisting of all sequences $\rho = (\rho_1, \rho_2, \dots)$, with entries $\rho_n \in \mathbb{C}$, which satisfy $\|\rho\|_\infty := \sup\{|\rho_n| : n \in \mathbb{N}\} < \infty$ is denoted by ℓ^∞. Show that ℓ^∞ is a Banach space with respect to $\|\cdot\|_\infty$ and that c_0 is a proper (i.e. $c_0 \neq \ell^\infty$) closed subspace of ℓ^∞. ∎

Let Ω be a topological space. The smallest σ-algebra on Ω containing all the open sets is called the *Borel σ-algebra* on Ω and is denoted by $Bo(\Omega)$.

Definition I.2. A complex measure $\nu : Bo(\Omega) \longrightarrow \mathbb{C}$, where Ω is a compact Hausdorff space, is called *regular* if for each $E \in Bo(\Omega)$ and $\varepsilon > 0$ there exists a compact set $K \subseteq \Omega$ and an open set $U \subseteq \Omega$ such that $K \subseteq E \subseteq U$ and $|\nu|(U \backslash K) < \varepsilon$. ∎

Theorem I.2. *Let Ω be a compact, topological Hausdorff space and $\Lambda : C(\Omega) \longrightarrow \mathbb{C}$ be an element of the dual space of $(C(\Omega), \|\cdot\|_\infty)$. Then there exists a regular complex measure $\mu : Bo(\Omega) \longrightarrow \mathbb{C}$ such that*

$$\langle \phi, \Lambda \rangle = \int_\Omega \phi \, d\mu, \qquad \phi \in C(\Omega).$$

Moreover, μ is unique in the sense that if $\nu : Bo(\Omega) \longrightarrow \mathbb{C}$ is another regular complex measure such that

$$\langle \phi, \Lambda \rangle = \int_\Omega \phi \, d\nu, \qquad \phi \in C(\Omega),$$

then $\mu = \nu$ (i.e. $\mu(E) = \nu(E)$ for all $E \in Bo(\Omega)$).

The above classical result is referred to as the *Riesz representation theorem*; see [38; Chapter 6], for example. It is important to note that if $\Lambda \in C(\Omega)'$ is represented by the regular measure $\mu : Bo(\Omega) \longrightarrow \mathbb{C}$ (i.e. $\langle \phi, \Lambda \rangle = \int_\Omega \phi \, d\mu$, for $\phi \in C(\Omega)$), then its dual norm

$$\|\Lambda\| = \sup\{|\langle \phi, \Lambda \rangle| : \phi \in C(\Omega), \ \|\phi\|_\infty \leq 1\}$$

equals the total variation $\|\mu\| := |\mu|(\Omega)$.

We have seen if X is a Banach space, then so is its dual space X' when equipped with the dual norm $\|\cdot\|_{X'}$. In Exercise 4 it is shown that $(c_0)' = \ell^1$ and $(\ell^1)' = \ell^\infty$. The Riesz representation theorem shows that $C(\Omega)'$ is the space of all regular Borel measures $\nu : Bo(\Omega) \longrightarrow \mathbb{C}$ equipped with the total variation norm $\|\nu\|$.

Definition I.3. Let X be a Banach space. The weakest topology on X which makes each element $\psi : X \longrightarrow \mathbb{C}$ of X' continuous is called the *weak topology* of X. A typical open neighbourhood of $x \in X$ for the weak topology has the form

$$\{y \in X : |\langle x, \psi \rangle - \langle y, \psi \rangle| < \varepsilon, \quad \psi \in \mathcal{F}\},$$

for some $\varepsilon > 0$ and some finite set $\mathcal{F} \subset X'$. In particular, a net of elements $\{x_\alpha\} \subseteq X$ converges to $x \in X$ for the weak topology if and only if

$$\lim_\alpha \langle x_\alpha, \psi \rangle = \langle x, \psi \rangle, \qquad \psi \in X'.$$

The weak topology on X will be denoted by $\sigma(X, X')$.   ∎

Example 4. Let $X = c_0$, in which case $X' = \ell^1$. Consider the sequence $\{e_n\}_{n=1}^\infty \subseteq X$, where e_n has 1 in the n-th co-ordinate and 0 elsewhere. Then $\{e_n\}_{n=1}^\infty$ converges to $0 \in X$ in the weak topology. Indeed, let $\xi = (\xi_1, \xi_2, \dots) \in X'$ in which case $\sum_{k=1}^\infty |\xi_k| < \infty$. Then

$$\lim_{n \to \infty} \langle e_n, \xi \rangle = \lim_{n \to \infty} \xi_n = 0 = \langle 0, \xi \rangle,$$

since the individual terms of an absolutely convergent series in $\mathbb{C}$ converge to 0. However, $\{e_n\}_{n=1}^\infty$ does *not* converge to 0 with respect to the norm of X since $\|e_n\|_\infty = 1$, for every $n \in \mathbb{N}$.   ∎

Exercise 5.[*] Show that if a sequence $\{x_n\}_{n=1}^\infty$ in a Banach space X converges in norm to $x \in X$, then $\{x_n\}_{n=1}^\infty$ also converges to x with respect to the weak topology. ∎

We record some basic facts about the weak topology. First, a subset $A \subset X$ is called *weakly bounded* if $|A|_\psi := \sup\{|\langle x, \psi \rangle| : \ x \in A\} < \infty$, for all $\psi \in X'$. It is called *norm bounded* if $\|A\| := \sup\{\|x\| : x \in A\} < \infty$. Recall that a subset A of a vector space is called *convex* if it has the property that $\lambda x + (1 - \lambda)y \in A$ whenever $x, y \in A$ and $\lambda \in [0, 1]$.

For the following basic facts about the weak topology we refer to [14; Chapter V].

Theorem I.3. *Let X be a Banach space.*

(a) *A subset of X is norm bounded if and only if it is weakly bounded .*

(b) *If A is any convex subset of X, then the closure of A with respect to the norm topology coincides with the closure of A with respect to the weak topology.*

(c) *A linear functional $\Lambda : X \longrightarrow \mathbb{C}$ is continuous with respect to the topology $\sigma(X, X')$ on X if and only if $\Lambda \in X'$.*

Exercise 6.[*] Let X be a Banach space and Y be a vector subspace of X. Show that the closure of Y in X with respect to the weak topology $\sigma(X, X')$ is the same as its closure in X with respect to the norm topology.   ∎

Definition I.4. Let X be a Banach space and $\sum_{n=1}^\infty x_n$ be a series of elements $x_n \in X$.

(i) The series is said to be *unconditionally norm convergent* if there exists $x \in X$ such that $\sum_{n=1}^\infty x_{\pi(n)} = x$, for all bijections $\pi : \mathbb{N} \longrightarrow \mathbb{N}$, that is, if $\lim_{N \to \infty} \|x - \sum_{n=1}^N x_{\pi(n)}\| = 0$, for all bijections $\pi : \mathbb{N} \to \mathbb{N}$.

(ii) The series is said to be *weakly subseries convergent* if each subseries $\sum_{k=1}^\infty x_{n_k}$ converges (to some element of X) in the weak topology.   ∎

The following remarkable result is due to W. Orlicz and B.J. Pettis, [8; Chapter 1, §4].

Theorem I.4. *Let X be a Banach space. Then a series $\sum_{n=1}^\infty x_n$ in X is unconditionally norm convergent whenever it is weakly subseries convergent.*

Exercise 7. A series $\sum_{n=1}^\infty x_n$ in a Banach space X is called *absolutely convergent* if

$$\sum_{n=1}^\infty \|x_n\| < \infty.$$

(a) Show that every absolutely convergent series is unconditionally norm convergent.

(b) Let X be the Hilbert space ℓ^2 of all sequences $\xi = (\xi_1, \xi_2, \dots)$, with $\xi_j \in \mathbb{C}$, which satisfy $\|\xi\|_2 := (\sum_{n=1}^{\infty} |\xi_n|^2)^{1/2} < \infty$. For each $n \in \mathbb{N}$, let $x_n \in X$ be the vector with $\frac{1}{n}$ in the n-th co-ordinate and 0 elsewhere. Show that $\sum_{n=1}^{\infty} x_n$ is unconditionally norm convergent, but *not* absolutely convergent. ∎

Definition I.5. The weakest topology on X' which makes each of the linear functionals $\langle x, \cdot \rangle : X' \longrightarrow \mathbb{C}$, $x \in X$, defined by $\psi \mapsto \langle x, \psi \rangle$, for $\psi \in X'$, continuous is called the *weak-star topology* of X' and is denoted by $\sigma(X', X)$. A typical open neighbourhood of $\psi \in X'$ has the form

$$\{\phi \in X' : |\langle x, \psi \rangle - \langle x, \phi \rangle| < \varepsilon, \quad x \in \mathcal{F}\},$$

for some $\varepsilon > 0$ and some finite set $\mathcal{F} \subset X$. In particular, a net of elements $\{\psi_\alpha\} \subseteq X'$ converges to $\psi \in X'$ for the weak-star topology if and only if

$$\lim_{\alpha} \langle x, \psi_\alpha \rangle = \langle x, \psi \rangle, \quad x \in X. \qquad ∎$$

Let X be a Banach space. Then X' determines the norm of X in that

$$(1) \qquad \|x\| = \sup\{|\langle x, x' \rangle| : x' \in X', \|x'\|_{X'} \le 1\}.$$

Now X' is itself a Banach space with respect to the dual norm

$$\|x'\|_{X'} = \sup\{|\langle x, x' \rangle| : x \in X, \|x\| \le 1\}$$

and hence, X' has again a continuous dual space $(X')'$, denoted simply by X'', consisting of all the linear functionals $\rho : X' \longrightarrow \mathbb{C}$ for which

$$(2) \qquad \|\rho\|_{X''} := \sup\{|\langle x', \rho \rangle| : x' \in X', \|x'\|_{X'} \le 1\} < \infty.$$

Clearly every element $x \in X$ defines an element $\hat{x} \in X''$ by

$$\hat{x} : x' \mapsto \langle x, x' \rangle, \quad x' \in X';$$

it is straight-forward to see from (1) and (2) that $\|\hat{x}\|_{X''} = \|x\|_X$. Hence, the map $x \mapsto \hat{x}$ from X into X'' is linear and *isometric*, meaning precisely that $\|x\|_X = \|\hat{x}\|_{X''}$ for all $x \in X$. It is called the *natural embedding* of X into X''. The space X'' is also called the *bidual* of X.

Definition I.6. Let X be a Banach space. If the natural embedding $x \mapsto \hat{x}$ of X into X'' is *surjective*, that is, it maps X *onto* the bidual X'', then X is called *reflexive*. ∎

Example 5. (a) Every Hilbert space is reflexive .

(b) Let $1 < p < \infty$ and ℓ^p denote the Banach space of all complex sequences $\xi = (\xi_1, \xi_2, \dots)$ with norm $\|\xi\|_p = (\sum_{n=1}^{\infty} |\xi_n|^p)^{1/p} < \infty$. If $1 < q < \infty$ is the unique number satisfying $\frac{1}{p} + \frac{1}{q} = 1$, then the dual Banach space of ℓ^p is ℓ^q, where each $\rho = (\rho_1, \rho_2, \dots) \in \ell^q$ acts on ℓ^p via the formula

$$\langle \xi, \rho \rangle := \sum_{n=1}^{\infty} \xi_n \rho_n, \quad \xi = (\xi_1, \xi_2, \dots) \in \ell^p.$$

Clearly each space ℓ^p is reflexive.

(c) The space c_0 is *not* reflexive , since $(c_0)' = \ell^1$ and $(\ell^1)' = \ell^\infty$ which contains c_0 as a *proper* closed subspace; see Exercise 4(c).   ∎

We now collect together a few basic facts needed later; parts (a)–(e) can be found in [14; Chapter V]. Part (f) can be found in [14; Chapter II, §3], for example; it is a consequence of the *Hahn-Banach theorem.*

Theorem I.5. *Let X be a Banach space.*

(a) *A subset A of X' is norm bounded if and only if it is weak-star bounded, that is,*

$$\sup\{|\langle x, x'\rangle| : x' \in A\} < \infty, \quad x \in X.$$

(b) *A subset A of X' is compact for the weak-star topology $\sigma(X', X)$ if and only if A is norm bounded and weak-star closed (Alaoglu's theorem).*

(c) *If $x \mapsto \hat{x}$ is the natural embedding of X into X'' , then $\hat{X} = \{\hat{x} : x \in X\}$ is $\sigma(X'', X')$-dense in X''.*

(d) *X is reflexive if and only if $\{x : \|x\| \leq 1\}$ is compact for the weak topology $\sigma(X, X')$.*

(e) *(Eberlein-Šmulian) A subset $A \subseteq X$ has the property that its closure in the $\sigma(X, X')$ topology is compact for the weak topology if and only if every sequence of elements from A has a subsequence which is $\sigma(X, X')$-convergent to some element of X.*

(f) *Let Y be a closed subspace of X and $y' \in Y'$. Then there exists $x' \in X'$ such that $\|x'\|_{X'} = \|y'\|_{Y'}$ and $\langle y, x'\rangle = \langle y, y'\rangle$ for all $y \in Y$.*

Exercise 8.   A Banach space X is called *weakly sequentially complete* if every sequence $\{x_n\}_{n=1}^\infty \subseteq X$ which is Cauchy for the weak topology $\sigma(X, X')$ converges to some $x \in X$ with respect to the weak topology.

(a) Show that every reflexive Banach space is weakly sequentially complete.

(b) Show that c_0 is *not* weakly sequentially complete .

(c) Note that ℓ^1 is weakly sequentially complete , but not reflexive; see [14; Chapter IV, §15], for example.   ∎

For the remainder of this chapter we concentrate on *vector measures* and, more specifically, on the theory of integration with respect to such measures.

Definition I.7.   Let X be a Banach space and (Ω, Σ) be a measurable space. A function $m : \Sigma \longrightarrow X$ is called a *finitely additive vector measure* if $m(\cup_{n=1}^k E_n) = \Sigma_{n=1}^k m(E_n)$ for all finite collections $\{E_n\}_{n=1}^k \subseteq \Sigma$ of pairwise disjoint sets. If, in addition, m satisfies

$$(3) \qquad\qquad\qquad m(\cup_{n=1}^\infty E_n) = \sum_{n=1}^\infty m(E_n),$$

for all sequences of pairwise disjoint sets $\{E_n\}_{n=1}^\infty \subseteq \Sigma$, where the series is convergent in the norm topology of X, then m is simply called a *vector measure.* A similar remark as for complex measures applies to show that the series $\sum_{n=1}^\infty m(E_n)$ is then necessarily unconditionally norm convergent in X. We also say that m is σ-additive on Σ.   ∎

Let $m : \Sigma \longrightarrow X$ be a finitely additive vector measure. For each $x' \in X'$ we define a set function $\langle m, x' \rangle : \Sigma \longrightarrow \mathbb{C}$ by $E \mapsto \langle m(E), x' \rangle$, for $E \in \Sigma$.

Proposition I.1. *Let X be a Banach space and (Ω, Σ) be a measurable space. Let $m : \Sigma \longrightarrow X$ be a finitely additive vector measure. Then m is a vector measure (i.e. m is σ-additive) if and only if $\langle m, x' \rangle : \Sigma \longrightarrow \mathbb{C}$ is a complex measure , for each $x' \in X'$.*

Proof. If m is σ-additive, then the continuity of x' and (3) imply that

$$\langle m, x' \rangle(\cup_{n=1}^{\infty} E_n) = \Big\langle \sum_{n=1}^{\infty} m(E_n), x' \Big\rangle = \sum_{n=1}^{\infty} \langle m(E_n), x' \rangle = \sum_{n=1}^{\infty} \langle m, x' \rangle(E_n)$$

whenever $\{E_n\}_{n=1}^{\infty} \subseteq \Sigma$ is a sequence of pairwise disjoint sets. Hence, $\langle m, x' \rangle$ is σ-additive.

Conversely, suppose that $\langle m, x' \rangle$ is σ-additive for each $x' \in X'$. Let $\{E_n\}_{n=1}^{\infty} \subseteq \Sigma$ be a pairwise disjoint sequence of sets. If $\{n_k\}_{k=1}^{\infty}$ is any increasing sequence of elements from $\mathbb{N}$, then the σ-additivity of $\langle m, x' \rangle$ implies that $\langle m(\cup_{k=1}^{\infty} E_{n_k}), x' \rangle = \sum_{k=1}^{\infty} \langle m(E_{n_k}), x' \rangle$, for each $x' \in X'$. This shows that the subseries $\sum_{k=1}^{\infty} m(E_{n_k})$ is weakly convergent to $m(\cup_{k=1}^{\infty} E_{n_k})$. By Theorem I.4 the series $\sum_{n=1}^{\infty} m(E_n)$ converges unconditionally in norm to $m(\cup_{n=1}^{\infty} E_n)$. This is precisely the requirement for m to be σ-additive in X. ∎

Corollary I.1.1. *Let $m : \Sigma \longrightarrow X$ be a vector measure on a σ-algebra Σ. Then its range $m(\Sigma) := \{m(E) : E \in \Sigma\}$ is a norm bounded subset of X.*

Proof. By Theorem I.3(a) it suffices to show that

$$(4) \qquad \sup\{|\langle m, x' \rangle(E)| : E \in \Sigma\} = \sup\{|\langle m(E), x' \rangle| : E \in \Sigma\} < \infty, \qquad x' \in X'.$$

But, by Proposition I.1 each $\langle m, x' \rangle$ is a complex measure and so Theorem I.1 implies that (4) is indeed finite, for each $x' \in X'$. ∎

It is important to note that Corollary I.1.1 *fails* to hold for finitely additive vector measures (in general), even when Σ is a σ-algebra and $X = \mathbb{C}$. For example, let $e_n, n \in \mathbb{N}$, be the standard unit vector in ℓ^{∞} with 1 at position n and 0 elsewhere. Extend $\{e_n : n \in \mathbb{N}\}$ to a Hamel basis of ℓ^{∞}. For $x \in \ell^{\infty}$, let $f_n(x)$ be the e_n-coordinate of x with respect to this Hamel basis. Then $f_n : \ell^{\infty} \longrightarrow \mathbb{C}$ is linear and $\{n : f_n(x) \neq 0\}$ is finite for each $x \in \ell^{\infty}$. For $E \in \Sigma := 2^{\mathbb{N}}$ define $m(E) = \sum_{n=1}^{\infty} f_n(\chi_E)$. Then $m : \Sigma \longrightarrow \mathbb{C}$ is a finitely additive measure with unbounded range $m(\Sigma)$ in $\mathbb{C}$ (since $m(\{1, 2, \ldots, n\}) = n$ for each $n \in \mathbb{N}$).

In view of this example, a finitely additive vector measure $m : \Sigma \longrightarrow X$ defined on an algebra of sets Σ is called *bounded* if its range $m(\Sigma)$ is a bounded subset of X. All finitely additive vector measures considered in these notes will be defined on either an algebra or σ-algebra of sets. Without any further qualification the phrase "$m : \Sigma \longrightarrow X$ is a vector measure" will mean that Σ is a σ-algebra and m is σ-additive. We point out that for any *bounded finitely additive* measure $\nu : \Sigma \longrightarrow \mathbb{C}$ defined on an *algebra* of sets Σ, all the properties (a)–(f) of Theorem I.1 remain valid (cf. [14; Chapter III]), except that $|\nu|$ in (b) is then only finitely additive.

Exercise 9. (a) Let $\sum_{n=1}^{\infty} x_n$ be a convergent series in a Banach space X. Show that $\lim_{n \to \infty} \|x_n\| = 0$.

(b) Let Σ be a σ-algebra of sets of a non-empty set and $m : \Sigma \longrightarrow X$ be finitely additive. Show that m is σ-additive if and only if $\lim_{n \to \infty} \|m(E_n)\| = 0$ whenever $\{E_n\}_{n=1}^{\infty} \subseteq \Sigma$ is a decreasing sequence with $\cap_{n=1}^{\infty} E_n = \emptyset$.   ■

Example 6. (a) Let $X := c_0$ and Σ be the σ-algebra of all subsets of $\Omega = \mathbb{N}$. Define $m : \Sigma \longrightarrow X$ by

$$m(E) = (\chi_E(1), \frac{1}{2}\chi_E(2), \frac{1}{3}\chi_E(3), \dots), \qquad E \in \Sigma.$$

To see that m is a vector measure , let $\xi = (\xi_1, \xi_2, \dots)$ belong to $X' = \ell^1$, in which case

(5) $$\langle m, \xi \rangle(E) = \sum_{n=1}^{\infty} \frac{\xi_n}{n} \chi_E(n) = \sum_{n=1}^{\infty} \frac{\xi_n}{n} \delta_n(E), \qquad E \in \Sigma,$$

where $\delta_n(E) := \chi_E(n)$, for $E \in \Sigma$, is the Dirac point measure at $n \in \Omega$. It is clear from (5) that $\langle m, \xi \rangle$ is a complex measure and so the conclusion follows from Proposition I.1.

(b) Let $X = \ell^{\infty}$ and (Ω, Σ) be as in part (a). Define $m(E) = (\chi_E(1), \chi_E(2), \dots)$, an element of ℓ^{∞}, for each $E \in \Sigma$. Then m is finitely additive , but not σ-additive . This follows from Exercise 9 applied to the sets $E_n = \{n, n+1, \dots\}$, for each $n \in \mathbb{N}$.   ■

Definition I.8. Let X be a Banach space and $m : \Sigma \longrightarrow X$ be a finitely additive vector measure defined on an algebra of sets Σ. The function $\|m\| : \Sigma \longrightarrow [0, \infty]$ defined by

$$\|m\|(E) = \sup\{|\langle m, x' \rangle|(E) : \ \|x'\| \leq 1, \quad x' \in X'\}, \quad E \in \Sigma,$$

is called the *semivariation* of m; here $|\langle m, x' \rangle|$ is the variation measure of the finitely additive complex measure $\langle m, x' \rangle$.   ■

The next result is one of the fundamental inequalities concerning vector measures.

Proposition I.2. *Let X be a Banach space and $m : \Sigma \longrightarrow X$ be a bounded finitely additive vector measure defined on an algebra of sets Σ. Then*

(6) $\sup\{\|m(H)\| : E \supseteq H \in \Sigma\} \leq \|m\|(E) \leq 4\sup\{\|m(H)\| : E \supseteq H \in \Sigma\}, \qquad E \in \Sigma.$

Proof. Fix $E \in \Sigma$. If $H \subseteq E$ and $H \in \Sigma$, then

$$\begin{aligned} \|m(H)\| &= \sup\{|\langle m(H), x' \rangle| : \|x'\| \leq 1\} \leq \sup\{|\langle m, x' \rangle|(H) : \|x'\| \leq 1\} \\ &\leq \sup\{|\langle m, x' \rangle|(E) : \|x'\| \leq 1\} = \|m\|(E), \end{aligned}$$

where the first inequality follows from Theorem I.1(c) and the second from Theorem I.1(e); see the discussion after Corollary I.1.1. This establishes the first inequality in (6).

Now let $x' \in X'$ satisfy $\|x'\| \leq 1$. Then, by Theorem I.1(f) and the discussion after Corollary I.1.1, we have that

$$|\langle m, x' \rangle|(E) \leq 4\sup\{|\langle m(H), x' \rangle| : E \supseteq H \in \Sigma\} \leq 4\sup\{\|m(H)\| : E \supseteq H \in \Sigma\}$$

as $\|m(H)\| = \sup\{|\langle m(H), z'\rangle| : z' \in X', \|z'\| \leq 1\}$. Accordingly,

(7) $$\sup\{|\langle m, x'\rangle|(E) : \|x'\| \leq 1\} \leq 4\sup\{\|m(H)\| : E \supseteq H \in \Sigma\}.$$

Since the left-hand-side of (7) is precisely $\|m\|(E)$ the second inequality in (6) follows. ∎

It is clear from Proposition I.2 and Corollary I.1.1 that $\|m\|(E)$ is *finite*, for each $E \in \Sigma$. We now determine some further basic properties of the semivariation of a vector measure.

Lemma I.1. *Let X be a Banach space, $m : \Sigma \longrightarrow X$ be a bounded finitely additive vector measure , $\|m\| : \Sigma \longrightarrow [0, \infty]$ be its semivariation , and $E, F \in \Sigma$.*

(a) $\|m\|(E) \leq \|m\|(F)$ *whenever $E \subseteq F$ (i.e. $\|m\|$ is monotone) .*

(b) $\|m\|(E \cup F) \leq \|m\|(E) + \|m\|(F)$, *that is, $\|m\|$ is subadditive .*

(c) $|\, \|m\|(E) - \|m\|(F)| \leq \|m\|(E\triangle F)$, *where $E\triangle F := (E\backslash F) \cup (F\backslash E)$.*

Proof. Both (a) and (b) are clear from the definition of $\|m\|$.

(c) We have to verify the two inequalities

$$\|m\|(E) \leq \|m\|(F) + \|m\|(E\triangle F)$$

and

$$\|m\|(F) \leq \|m\|(E) + \|m\|(E\triangle F).$$

Since the roles of E and F are symmetric it suffices to establish the first inequality. Write $E = (E \cap F) \cup (E\backslash F)$ and note that the two sets in the union are disjoint. Then

$$\|m\|(E) \leq \|m\|(E \cap F) + \|m\|(E\backslash F) \leq \|m\|(F) + \|m\|(E\triangle F),$$

as required, where the first inequality follows from part (b) and the second from part (a) as $(E \cap F) \subseteq F$ and $E\backslash F \subseteq E\triangle F$. ∎

A sequence of sets $\{E_n\}_{n=1}^{\infty}$ is called *convergent* if $\cup_{n=1}^{\infty}(\cap_{k=n}^{\infty} E_k) = \cap_{n=1}^{\infty}(\cup_{k=n}^{\infty} E_k)$; this set is then denoted by $\lim_n E_n$.

Proposition I.3. *Let X be a Banach space, $m : \Sigma \longrightarrow X$ be a vector measure and $\{E_n\}_{n=1}^{\infty} \subseteq \Sigma$ be a convergent sequence of sets. Then*

$$\|m\|(\lim_n E_n) = \lim_{n\to\infty} \|m\|(E_n).$$

Proof. We begin with a special case. Namely, suppose that the sequence $\{E_n\}_{n=1}^{\infty}$ is decreasing and $\cap_{n=1}^{\infty} E_n = \emptyset$. Assume that $\lim_{n\to\infty} \|m\|(E_n) \neq 0$. Then there exists $\varepsilon > 0$ such that $\|m\|(E_n) > \varepsilon$ for all $n \in \mathbb{N}$. Let $n_1 = 1$. Then there must exist $x' \in X'$ with $\|x'\| \leq 1$ such that $|\langle m, x'\rangle|(E_{n_1}) > \varepsilon$. Since $|\langle m, x'\rangle|(E_n) \downarrow 0$, there is $n_2 > n_1$ such that $|\langle m, x'\rangle|(E_{n_2}) < \frac{1}{2}\varepsilon$. Then

$$4\sup\{\|m(F)\| : F \in \Sigma, \ F \subseteq (E_{n_1}\backslash E_{n_2})\} \geq |\langle m, x'\rangle|(E_{n_1}\backslash E_{n_2})$$

by Theorem I.1(f) and Proposition I.2. But,

$$|\langle m, x' \rangle|(E_{n_1} \backslash E_{n_2}) = |\langle m, x' \rangle|(E_{n_1}) - |\langle m, x' \rangle|(E_{n_2}) > \frac{1}{2}\varepsilon$$

since $|\langle m, x' \rangle|(E_{n_1}) > \varepsilon$ and $|\langle m, x' \rangle|(E_{n_2}) < \frac{1}{2}\varepsilon$. Accordingly, there exists $F_1 \in \Sigma$ with $F_1 \subseteq (E_{n_1} \backslash E_{n_2})$ such that $\|m(F_1)\| > \frac{1}{8}\varepsilon$.

Again, since $\|m\|(E_{n_2}) > \varepsilon$, there exists $z' \in X'$ with $\|z'\| \leq 1$ such that $|\langle m, z' \rangle|(E_{n_2}) > \varepsilon$. Since $|\langle m, z' \rangle|(E_n) \downarrow 0$, there exists $n_3 > n_2$ such that $|\langle m, z' \rangle|(E_{n_3}) < \frac{1}{2}\varepsilon$. By the same estimates as above

$$4 \sup\{\|m(F)\| : F \in \Sigma, \ F \subseteq (E_{n_2} \backslash E_{n_3})\} > \frac{1}{2}\varepsilon$$

from which it follows that there exists $F_2 \in \Sigma$ with $F_2 \subseteq (E_{n_2} \backslash E_{n_3})$ such that $\|m(F_2)\| > \frac{1}{8}\varepsilon$. Continuing inductively gives an increasing sequence $\{n_k\}_{k=1}^{\infty} \subseteq \mathbb{N}$ and a sequence of sets $\{F_k\}_{k=1}^{\infty} \subseteq \Sigma$ with $F_k \subseteq (E_{n_k} \backslash E_{n_{k+1}})$ such that $\|m(F_k)\| > \frac{1}{8}\varepsilon$, for all $k \in \mathbb{N}$. Since the sets in $\{F_k\}_{k=1}^{\infty}$ are pairwise disjoint this contradicts the σ-additivity of m. So, the statement of the proposition is proved for decreasing sequences $E_n \downarrow \emptyset$.

Suppose now that $\{E_n\}_{n=1}^{\infty} \subseteq \Sigma$ is convergent with limit E. Since $E \triangle E_n \subseteq \cup_{k=n}^{\infty} E \triangle E_k$, for all $n \in \mathbb{N}$, it follows from Lemma I.1 that

$$|\ \|m\|(E) - \|m\|(E_n)| \leq \|m\|(E \triangle E_n) \leq \|m\|(\cup_{k=n}^{\infty} E \triangle E_k) =$$
$$= \|m\|([\cup_{k=n}^{\infty} E \backslash E_k] \cup [\cup_{k=n}^{\infty} E_k \backslash E]) \leq \|m\|(\cup_{k=n}^{\infty} E_k \backslash E) + \|m\|(\cup_{k=n}^{\infty} E \backslash E_k)$$

But, the sequence $\{\cup_{k=n}^{\infty} E \backslash E_k\}_{n=1}^{\infty}$ decreases to $\emptyset$ since

$$\cap_{n=1}^{\infty}(\cup_{k=n}^{\infty} E \backslash E_k) = \cap_{n=1}^{\infty}(E \backslash \cap_{k=n}^{\infty} E_k) = E \backslash \cup_{n=1}^{\infty} (\cap_{k=n}^{\infty} E_k) = E \backslash E = \emptyset.$$

The same is true of $\{\cup_{k=n}^{\infty} E_k \backslash E\}_{n=1}^{\infty}$. So, by the special case proved above, both terms in the sum on the right-hand-side of the previous inequality converge to 0 as $n \to \infty$. It follows that $\|m\|(E) = \lim_{n \to \infty} \|m\|(E_n)$. ∎

Let $\nu : \Sigma \longrightarrow \mathbb{C}$ be a complex measure . Then a set $E \in \Sigma$ is said to be ν-null if $|\nu|(E) = 0$. By Theorem I.1(c) and the definition of the variation measure $|\nu|$ this is equivalent to the statement that $\nu(F) = 0$ for every $F \in \Sigma$ with $F \subseteq E$.

Suppose that $\mu : \Sigma \longrightarrow [0, \infty)$ is a non-negative measure. We say that ν is *absolutely continuous* with respect to μ, written as $\nu \ll \mu$, if every μ-null set is also a ν-null set. It is known (combine Theorem I.1(f) with [38; Theorem 6.11]) that $\nu \ll \mu$ is equivalent to the requirement that for every $\varepsilon > 0$ there is a $\delta > 0$ such that $|\nu|(E) < \varepsilon$, for all $E \in \Sigma$ satisfying $\mu(E) < \delta$.

The following result, known as the *Radon-Nikodym theorem*, is one of the most important facts in measure theory; see [38; Chapter 6], for example. Recall that a measure $\mu : \Sigma \longrightarrow [0, \infty]$ is called σ-*finite* if Ω is a union of countably many sets $\Omega_n \in \Sigma$, for $n \in \mathbb{N}$, such that $\mu(\Omega_n) < \infty$ for each $n \in \mathbb{N}$.

Theorem I.6. *Let* (Ω, Σ) *be a measurable space . Let* $\nu : \Sigma \longrightarrow \mathbb{C}$ *be a complex measure and* $\mu : \Sigma \longrightarrow [0, \infty]$ *be a non-negative, σ-finite measure. If* $\nu \ll \mu$, *then there exists a unique element* $h \in L^1(\mu)$ *such that*

$$\nu(E) = \int_E h \, d\mu, \qquad E \in \Sigma.$$

Moreover, the variation measure $|\nu| : \Sigma \longrightarrow [0, \infty)$ *of* ν *is given by*

$$|\nu|(E) = \int_E |h| \, d\mu, \qquad E \in \Sigma.$$

Suppose now that $m : \Sigma \longrightarrow X$ is a vector measure . Then a set $E \in \Sigma$ is said to be *m-null* if $m(F) = 0$ for every $F \in \Sigma$ with $F \subseteq E$. By Proposition I.2 this is equivalent to requiring $\|m\|(E) = 0$. Given a finite measure $\mu : \Sigma \longrightarrow [0, \infty)$ we say that m is *absolutely continuous* with respect to μ, also denoted by $m \ll \mu$, if for every $\varepsilon > 0$ there is a $\delta > 0$ such that $\|m\|(E) < \varepsilon$, for all $E \in \Sigma$ satisfying $\mu(E) < \delta$.

Theorem I.7. *Let* X *be a Banach space,* $m : \Sigma \longrightarrow X$ *be a vector measure and* $\mu : \Sigma \longrightarrow [0, \infty)$ *be a finite, non-negative measure. Then* $m \ll \mu$ *if and only if* $m(E) = 0$ *whenever* $E \in \Sigma$ *satisfies* $\mu(E) = 0$.

Since $\mu(E) = 0$ implies $\mu(F) = 0$ for every $F \in \Sigma$ with $F \subseteq E$, it follows that $m(F) = 0$ for all such sets F and hence, that $\|m\|(E) = 0$. Accordingly, $m \ll \mu$ if and only if every μ-null set is also m-null . Theorem I.7, due to B.J. Pettis , can be found in [8; p.10].

Concerning the *existence* of measures μ for which $m \ll \mu$, we record the following fundamental result, due to R.G. Bartle , N. Dunford and J.T. Schwartz in 1955; see [8; p.14], for example.

Theorem I.8. *Let* X *be a Banach space and* $m : \Sigma \longrightarrow X$ *be a vector measure . Then there exists a finite measure* $\mu : \Sigma \longrightarrow [0, \infty)$ *such that* $m \ll \mu$. *Moreover,* μ *can be chosen to satisfy* $0 \le \mu(E) \le \|m\|(E)$, *for each* $E \in \Sigma$.

Let $m : \Sigma \longrightarrow X$ be a bounded finitely additive vector measure defined on an *algebra* of sets Σ. If $f = \sum_{j=1}^{n} \alpha_j \chi_{E(j)}$ is a Σ-*simple function* on Ω, where $\alpha_j \in \mathbb{C}$ and $E(j)$, for $1 \le j \le n$, are *pairwise disjoint* members from Σ, then we define

$$(8) \qquad \int_E f \, dm := \sum_{j=1}^{n} \alpha_j m(E \cap E(j)), \qquad E \in \Sigma.$$

Note that the standard argument used for finitely additive complex measures also applies to show that the integral in (8) is well defined, that is, it is independent of the particular representation used for f. Let $\beta = \sup\{|f(w)| : w \in \Omega\} = \max\{|\alpha_j| : 1 \leq j \leq n\}$. Then

$$
\begin{aligned}
\left\| \int_E f\, dm \right\| &= \beta \left\| \sum_{j=1}^{n} \beta^{-1} \alpha_j m(E(j) \cap E) \right\| \\
&= \beta \sup\{ |\langle \sum_{j=1}^{n} \beta^{-1} \alpha_j m(E(j) \cap E), x' \rangle | : x' \in X', \|x'\| \leq 1 \} \\
&\leq \beta \sup\{ \sum_{j=1}^{n} \beta^{-1} |\alpha_j| \cdot |\langle m(E(j) \cap E), x' \rangle| : \|x'\| \leq 1 \} \\
&\leq \beta \sup\{ \sum_{j=1}^{n} |\langle m(E(j) \cap E), x' \rangle| : \|x'\| \leq 1 \}, \quad \text{as } \beta^{-1}|\alpha_j| \leq 1, \\
&\leq \beta \sup\{ \sum_{j=1}^{n} |\langle m, x' \rangle| (E(j) \cap E) : \|x'\| \leq 1 \}, \quad \text{by Theorem I.1,} \\
&= \beta \sup\{ |\langle m, x' \rangle| (E \cap (\cup_{j=1}^{n} E(j)) : \|x'\| \leq 1 \}, \quad \text{by disjointness of } \{E(j)\}_{j=1}^{n}, \\
&\leq \beta \sup\{ |\langle m, x' \rangle| (E) : \|x'\| \leq 1 \}, \quad \text{by Theorem I.1,} \\
&= \beta \|m\|(E),
\end{aligned}
$$

for each $E \in \Sigma$, where the references to Theorem I.1 are in combination with the discussion after Corollary I.1.1. So, for each Σ-simple function $f : \Omega \longrightarrow \mathbb{C}$ we have

$$
(9) \qquad \left\| \int_E f\, dm \right\| \leq \|m\|(E) \cdot \|f\|_\infty, \qquad E \in \Sigma.
$$

Let $B^\infty(\Sigma)$ denote the *closure* of the vector space of all Σ-simple functions, formed in the Banach space of all bounded functions $f : \Omega \longrightarrow \mathbb{C}$ which are measurable with respect to the σ-algebra generated by Σ and equipped with the norm $\|f\|_\infty = \sup\{|f(w)| : w \in \Omega\}$, [38; p.16]. Since the Σ-simple functions are clearly dense in $B^\infty(\Sigma)$, it follows from (9) that the linear map $f \mapsto \int_E f\, dm$ (for fixed $E \in \Sigma$) can be extended to a continuous linear map (with norm at most $\|m\|(E)$) on all of $B^\infty(\Sigma)$; this is a special case of Theorem 18 in [14; p.55]. We again write $\int_E f\, dm$ for each $f \in B^\infty(\Sigma)$ and note that (9) still holds for all $f \in B^\infty(\Sigma)$. Moreover, for each $f \in B^\infty(\Sigma)$, we have

$$
(10) \qquad \langle \int_E f\, dm, x' \rangle = \int_E f\, d\langle m, x' \rangle, \qquad x' \in X'.
$$

This is immediate from (8) if f is a Σ-simple function. The case for a general function $f \in B^\infty(\Sigma)$ then follows by approximating f in $B^\infty(\Sigma)$ by Σ-simple functions, applying (9) to the bounded finitely additive measure $\langle m, x' \rangle$ in place of m, for a fixed $x' \in X'$, and using the fact that x' is linear and continuous.

We now wish to present an elegant application of the Bartle -Dunford-Schwartz theorem (c.f. Theorem I.8.), which is a significant refinement of Corollary I.1.1 in non-reflexive spaces.
Proposition I.4. *Let X be a Banach space and $m : \Sigma \longrightarrow X$ be a vector measure . Then its range $m(\Sigma)$ is a relatively weakly compact subset of X. That is, the closure in X of $m(\Sigma)$ with respect to the weak topology $\sigma(X, X')$ is weakly compact.*

Proof. By Theorem I.8 there is a finite measure $\mu : \Sigma \longrightarrow [0, \infty)$ such that $m \ll \mu$. Recall that $L^\infty(\mu)$ is the space of (equivalence classes of) Σ-measurable functions f such that

$$\|f\|_{L^\infty} := \inf\{\sup\{|f(w)| : w \in E\} : E \in \Sigma, \mu(E^c) = 0\} < \infty.$$

So, given $f \in L^\infty(\mu)$ there is a μ-null set $N(f) \in \Sigma$ such that $f\chi_{N(f)^c} \in B^\infty(\Sigma)$ and $\|f\|_{L^\infty} = \|f\chi_{N(f)^c}\|_\infty$. Since $m \ll \mu$ the set $N(f)$ is also m-null and so we have a well defined linear map $T : L^\infty(\mu) \longrightarrow X$ given by $Tf = \int_\Omega f\chi_{N(f)^c} \, dm$. For simplicity of notation we simply write $Tf = \int_\Omega f \, dm$, for $f \in L^\infty(\mu)$, since if $g = f$ up to a μ-null set (hence, also an m-null set) for some $g \in B^\infty(\Sigma)$, then $\int_\Omega g \, dm = \int_\Omega f\chi_{N(f)^c} \, dm$.

For each $x' \in X'$ we have $\langle m, x' \rangle \ll \mu$ and so the Radon-Nikodym theorem (c.f. Theorem I.6) guarantees the existence of $g_{x'} \in L^1(\mu)$ such that

$$\langle m(E), x' \rangle = \int_E g_{x'} \, d\mu, \qquad E \in \Sigma.$$

It then follows from (10) that

(11)
$$\langle Tf, x' \rangle = \int_\Omega f g_{x'} \, d\mu, \qquad f \in L^\infty(\mu), \, x' \in X'.$$

It is known that $(L^1(\mu))' = L^\infty(\mu)$, [38; Theorem 6.16]. So, let $\{f_\alpha\} \subseteq L^\infty(\mu)$ be a net which converges in the weak-star topology $\sigma(L^\infty(\mu), L^1(\mu))$ to some $h \in L^\infty(\mu)$. Fix $x' \in X'$. It follows from (11) that

$$\lim_\alpha \langle Tf_\alpha, x' \rangle = \lim_\alpha \int_\Omega f_\alpha g_{x'} \, d\mu = \int_\Omega h g_{x'} \, d\mu = \langle Th, x' \rangle.$$

Since $x' \in X'$ is arbitrary this shows that the net $\{Tf_\alpha\} \subseteq X$ converges to $Th \in X$ with respect to the weak topology $\sigma(X, X')$. So, T is continuous from $(L^\infty(\mu), \sigma(L^\infty(\mu), L^1(\mu)))$ into $(X, \sigma(X, X'))$. By Alaoglu's theorem (c.f. Theorem I.5(b)) the set $D = \{f \in L^\infty(\mu) : \|f\|_{L^\infty} \leq 1\}$ is $\sigma(L^\infty(\mu), L^1(\mu))$-compact. Since the continuous image of a compact set is again compact, [37; Chapter 9], it follows that $T(D)$ is compact in X for the weak topology . But, since $\|\chi_E\|_{L^\infty} \leq 1$ for each $E \in \Sigma$, we have $m(\Sigma) = \{T\chi_E : E \in \Sigma\} \subseteq T(D)$. This shows that $m(\Sigma)$ is relatively weakly compact. ∎

It is now time to turn to the theory of *integration* with respect to a vector measure .

Definition I.9. Let $m : \Sigma \longrightarrow X$ be a vector measure. Then a Σ-measurable function $f : \Omega \longrightarrow \mathbb{C}$ is called m-*integrable* if $f \in L^1(\langle m, x' \rangle)$, for each $x' \in X'$ (i.e. $\int_\Omega |f| \, d|\langle m, x' \rangle| < \infty$ for each $x' \in X'$) and if, for each $E \in \Sigma$, there exists a vector in X (necessarily unique), denoted by $\int_E f \, dm$, satisfying (10) for each $x' \in X'$. ∎

The vector space of all m-integrable functions is denoted by $L(m)$. Since elements of $B^\infty(\Sigma)$ are integrable with respect to *every* complex measure $\nu : \Sigma \longrightarrow \mathbb{C}$ it is clear from the discussion after Theorem I.8 and the formula (10) that $B^\infty(\Sigma) \subseteq L(m)$. It is also clear from Definition I.9 that if $f \in L(m)$, then $f\chi_E \in L(m)$ for all $E \in \Sigma$. So, given a $\mathbb{R}$-valued function $f \in L(m)$ it follows that both $f\chi_{\Omega(f)^+} = f \vee 0 := f^+$ and $f^- := (-f)^+ = (-f)\chi_{\Omega(f)^-}$ belong to $L(m)$, where $\Omega(f)^+ := \{w \in \Omega : f(w) \geq 0\}$ and $\Omega(f)^- := \{w \in \Omega : f(w) < 0\}$. Hence, also $|f| = f^+ + f^-$ belongs to $L(m)$. To deduce the same fact for $\mathbb{C}$-valued functions f requires an additional result.

Given $f \in L(m)$, define $m_f : \Sigma \longrightarrow X$ by $m_f(E) = \int_E f\,dm$, for $E \in \Sigma$. We call m_f the *indefinite integral* of f with respect to m.

Lemma I.2. *Let X be a Banach space, $m : \Sigma \longrightarrow X$ be a vector measure and $f \in L(m)$. Then $m_f : \Sigma \longrightarrow X$ is also a vector measure*

Proof. Let $x' \in X'$. Then

$$\langle m_f(E), x' \rangle = \langle \int_E f\,dm, x' \rangle = \int_E f\,d\langle m, x' \rangle, \qquad E \in \Sigma.$$

Since $f \in L^1(\langle m, x' \rangle)$, it is known from standard scalar-valued measure theory that $E \mapsto \int_E f\,d\langle m, x' \rangle$ is σ-additive, that is, it is a complex measure . So, $\langle m_f, x' \rangle$ is a complex measure, for each $x' \in X'$, and the conclusion follows from Proposition I.1. ∎

Lemma I.3. *Let X be a Banach space, $m : \Sigma \longrightarrow X$ be a vector measure and $f \in L(m)$. Then also $|f| \in L(m)$.*

Proof. Let $E = f^{-1}(\{0\}) = |f|^{-1}(\{0\})$, in which case $E \in \Sigma$. Then define $h := (\frac{|f|}{f})\chi_{E^c}$ and note that $h \in B^\infty(\Sigma)$ as $\|h\|_\infty \leq 1$. Clearly $|f| \in L^1(\langle m, x' \rangle)$ for all $x' \in X'$, as $L^1(\nu)$ has the property that $\phi \in L^1(\nu)$ implies $|\phi| \in L^1(\nu)$ for any complex measure ν. Since $h \in B^\infty(\Sigma)$, we have seen that $h \in L(m_f)$, as m_f is σ-additive by Lemma I.2. So, let $x_F := \int_F h\,dm_f$, for $F \in \Sigma$. Then, for $x' \in X'$, we have

$$\langle x_F, x' \rangle = \langle \int_F h\,dm_f, x' \rangle = \int_F h\,d\langle m_f, x' \rangle = \int_F hf\,d\langle m, x' \rangle = \int_F |f|\,d\langle m, x' \rangle.$$

This shows that $|f|$ is m-integrable and $\int_F f\,dm = x_F$, for each $F \in \Sigma$. ∎

A slight (but useful) extension of Lemma I.3 is the following result.

Lemma I.4. *Let X be a Banach space, $m : \Sigma \longrightarrow X$ be a vector measure and $g \geq 0$ be an m-integrable function. If $f : \Omega \longrightarrow \mathbb{C}$ is a Σ-measurable function such that $|f(w)| \leq g(w)$, for $w \in \Omega$, then $f \in L(m)$.*

Proof. Let $E = g^{-1}(\{0\})$ and note that $f^{-1}(\{0\}) \supseteq E$. Define $h := (\frac{f}{g})\chi_{E^c}$ and note that $h \in B^\infty(\Sigma)$. With $x_F := \int_F h\,dm_g$, for each $F \in \Sigma$, the argument of the proof of Lemma I.3 can be repeated. ∎

The following result summarizes the previous few facts.

Proposition I.5. *Let (Ω, Σ) be a measurable space, X be a Banach space and $m : \Sigma \longrightarrow X$ be a vector measure. Then $L(m)$ is a vector space and a (complex) vector lattice in the sense that;*

(a) $f \in L(m)$ *implies that* $|f| \in L(m)$.

(b) $0 \leq g \in L(m)$ *and f is Σ-measurable with $|f| \leq g$ implies that $f \in L(m)$.*

(c) $B^\infty(\Sigma) \subseteq L(m)$.

Remark. If $\nu : \Sigma \longrightarrow \mathbb{C}$ is a complex measure and $f \in L^1(\nu)$, then the measure $\nu_f : E \mapsto \int_E f \, d\nu$ has variation measure given by $|\nu_f|(E) = \int_E |f| \, d|\nu|$, for $E \in \Sigma$; this follows from [38; Theorems 6.12 & 6.13], for example. Suppose now that $m : \Sigma \longrightarrow X$ is a vector measure and $f \in L(m)$. Then we have

$$(12) \qquad \|m_f\|(E) = \|m_{|f|}\|(E), \quad E \in \Sigma.$$

Indeed, by the previous comment about complex measures and the definition of m_f, we have

$$
\begin{aligned}
\|m_f\|(E) &= \sup\{|\langle m_f, x'\rangle|(E) : \|x'\| \le 1\} \\
&= \sup\{\int_E |f| \, d|\langle m, x'\rangle| : \|x'\| \le 1\} \\
&= \sup\{|\langle m_{|f|}, x'\rangle|(E) : \|x'\| \le 1\} = \|m_{|f|}\|(E),
\end{aligned}
$$

for each $E \in \Sigma$. A similar calculation shows that if f and g are m-integrable functions satisfying $0 \le f \le g$, then

$$(13) \qquad \|m_f\|(E) \le \|m_g\|(E), \quad E \in \Sigma. \qquad \blacksquare$$

The following result, known as the *dominated convergence theorem* for vector measures will be an important tool in the sequel.

Theorem I.9. *Let X be a Banach space, $m : \Sigma \longrightarrow X$ be a vector measure and $0 \le g \in L(m)$. Let $f_n : \Omega \longrightarrow \mathbb{C}$, for $n \in \mathbb{N}$, be a sequence of Σ-measurable functions such that*

(i) $\lim_{n \to \infty} f_n(w) = f(w)$ *exists, for each $w \in \Omega$, and*

(ii) $|f_n| \le g$, *for $n = 1, 2, \ldots$*

Then $f \in L(m)$ and $\|m_{(f-f_n)}\|(\Omega) = \|m_f - m_{f_n}\|(\Omega) \longrightarrow 0$ as $n \to \infty$.

Proof. Fix $\varepsilon > 0$ and define $E_n = \{w \in \Omega : |f(w) - f_n(w)| \ge \varepsilon\}$, in which case $\lim_n E_n = \emptyset$. Fix $E \in \Sigma$. For each $x' \in X'$ satisfying $\|x'\| \le 1$ we know from the dominated convergence theorem for scalar measures that $f \in L^1(\langle m, x'\rangle)$. Moreover, for each $n \in \mathbb{N}$,

$$
\begin{aligned}
\left| \int_E (f - f_n) \, d\langle m, x'\rangle \right| &\le \int_E |f - f_n| \, d|\langle m, x'\rangle| \\
&= \int_{E \setminus E_n} |f - f_n| \, d|\langle m, x'\rangle| + \int_{E \cap E_n} |f - f_n| \, d|\langle m, x'\rangle| \\
&\le \varepsilon \|m\|(E \setminus E_n) + \int_{E \cap E_n} |f - f_n| \, d|\langle m, x'\rangle|.
\end{aligned}
$$

But, for $w \in E \cap E_n$ we have that

$$|(f - f_n)(w)| = \lim_{k \to \infty} |f_k(w) - f_n(w)| \le 2g(w)$$

and so, as $\|x'\| \leq 1$, that

$$\int_{E\cap E_n} |f - f_n|\, d|\langle m, x'\rangle| \leq 2\int_{E\cap E_n} g\, d|\langle m, x'\rangle| \leq 2\|m_g\|(E\cap E_n), \qquad n \in \mathbf{N}.$$

Hence, whenever $\|x'\| \leq 1$, Lemma I.1(a) applied to both m and m_g yields

$$\left| \int_E (f - f_n)\, d\langle m, x'\rangle \right| \leq \varepsilon\|m\|(E\backslash E_n) + 2\|m_g\|(E\cap E_n)$$
$$\leq \varepsilon\|m\|(\Omega) + 2\|m_g\|(E_n), \qquad n \in \mathbf{N}.$$

Take the supremum of the left-hand-side, with respect to $\|x'\| \leq 1$, gives

$$\left\| \int_E (f - f_n)\, dm \right\| \leq \varepsilon\|m\|(\Omega) + 2\|m_g\|(E_n), \qquad n \in \mathbf{N}.$$

It follows, for each $E \in \Sigma$, that

$$(14) \qquad \left\| \int_E f_k\, dm - \int_E f_n\, dm \right\| \leq \left\| \int_E (f - f_k)\, dm \right\| + \left\| \int_E (f - f_n)\, dm \right\|$$
$$\leq 2\varepsilon\|m\|(\Omega) + 2\|m_g\|(E_n) + 2\|m_g\|(E_k), \qquad k, n \in \mathbf{N}.$$

Since $\lim_n E_n = \emptyset$, we have from Proposition I.3 applied to the vector measure m_g that

$$\lim_{n\to\infty} \|m_g\|(E_n) = \|m_g\|(\lim_n E_n) = \|m_g\|(\emptyset) = 0.$$

So, (14) shows (as $\{\|m_g\|(E_n)\}_{n=1}^{\infty}$ is Cauchy in $[0,\infty)$) that $\{\int_E f_n\, dm\}_{n=1}^{\infty}$ is Cauchy in X, uniformly with respect to $E \in \Sigma$. By completeness of X there is $x_E \in X$ such that $\|\int_E f_n\, dm - x_E\| \longrightarrow 0$ as $n \to \infty$.

By (i) and (ii), the dominated convergence theorem for scalar-valued measures, and the fact that $g \in L^1(\langle m, x'\rangle)$, (see Definition I.9), we have $f \in L^1(\langle m, x'\rangle)$, for all $x' \in X'$. Moreover, defining $\int_E f\, dm := x_E$, gives

$$\langle x_E, x'\rangle = \lim_{n\to\infty} \left\langle \int_E f_n\, dm, x'\right\rangle = \lim_{n\to\infty} \int_E f_n\, d\langle m, x'\rangle = \int_E f\, d\langle m, x'\rangle.$$

This shows that indeed $f \in L(m)$.

Fix $\varepsilon > 0$ again. We have seen that there is $N > 0$ such that

$$\sup_{E\in\Sigma} \left\| \int_E f_k\, dm - \int_E f_n\, dm \right\| \leq \varepsilon/4, \qquad k, n \geq N.$$

Fix $F \in \Sigma$. Then for $n \geq N$ we have (n fixed)

$$\left\| \int_F f\, dm - \int_F f_n\, dm \right\| = \lim_{k\to\infty} \left\| \int_F f_k\, dm - \int_F f_n\, dm \right\| \leq \varepsilon/4.$$

Accordingly,

$$\sup_{F \in \Sigma} \left\| \int_F f \, dm - \int_F f_n \, dm \right\| \le \varepsilon/4, \qquad n \ge N.$$

Then, for every $n \ge N$, we have by Proposition I.2 that

$$\|m_f - m_{f_n}\|(\Omega) \le 4 \sup_{F \in \Sigma} \left\| \int_F f \, dm - \int_F f_n \, dm \right\| \le \varepsilon.$$

This shows that $\|m_f - m_{f_n}\|(\Omega) \longrightarrow 0$ as $n \to \infty$. ∎

Exercise 10. Let X be a Banach space which is weakly sequentially complete .

(a) Suppose $\sum_{n=1}^{\infty} x_n$ is a (formal) series in X such that $\sum_{n=1}^{\infty} |\langle x_n, x' \rangle| < \infty$ for each $x' \in X'$. Show that there exists $x \in X$ such that $\sum_{n=1}^{\infty} x_n$ is unconditionally (norm) convergent to x.

(b) Let X be as in part (a), $m : \Sigma \longrightarrow X$ be a vector measure and f be a $\mathbb{C}$-valued, Σ-measurable function such that $\int_{\Omega} |f| \, d|\langle m, x' \rangle| < \infty$, for every $x' \in X'$. Show that f is m-integrable. ∎

Definition I.10. Let $m : \Sigma \longrightarrow X$ be a vector measure . A function $f \in L(m)$ is called m-null if m_f is the zero vector measure , that is, $\int_E f \, dm = 0$ for all $E \in \Sigma$. By Proposition I.2 this is equivalent to $\|m_f\|(\Omega) = 0$. ∎

Two functions $f, g \in L(m)$ are called m-equivalent if $|f - g|$ is m-null. This is an equivalence relation on $L(m)$; see [37; Chapter 1, §7] for the definition of an equivalence relation. The equivalence class of $f \in L(m)$ is denoted by $[f]_m := \{g \in L(m) : f \text{ is } m\text{-equivalent to } g\}$. The operations $[f]_m + [g]_m := [f + g]_m$ and $\alpha [f]_m := [\alpha f]_m$, for $\alpha \in \mathbb{C}$ and $f, g \in L(m)$, are well defined and the *quotient space* of $L(m)$ with respect to this equivalence relation, denoted by $L^1(m)$, is a vector space . For the definition of the quotient space we again refer to [37; Chapter 1, §7], for example. Moreover, we can define a *norm* in $L^1(m)$ by

$$\| [f]_m \|_1 = \|m_f\|(\Omega), \qquad [f]_m \in L^1(m);$$

see [37; p.183], for example. Since there is no confusion likely, elements $[f]_m \in L^1(m)$ will also be denoted simply as $f \in L^1(m)$.

Exercise 11. Let $X = \ell^1$ and Σ be the σ-algebra of all subsets of $\mathbb{N}$. Define $m : \Sigma \longrightarrow X$ by

$$m(E) = (\chi_E(1), 2^{-2}\chi_E(2), 3^{-2}\chi_E(3), \dots), \qquad E \in \Sigma.$$

(a) Establish that m is a vector measure .

(b) Show that a function $f : \mathbb{N} \to \mathbb{C}$ is m-integrable if and only if $\sum_{n=1}^{\infty} n^{-2} |f(n)| < \infty$ in which case

$$\int_E f \, dm = (\chi_E(1) f(1), 2^{-2}\chi_E(2) f(2), 3^{-2}\chi_E(3) f(3), \dots), \qquad E \in \Sigma,$$

and $\| [f]_m \|_1 = \sum_{n=1}^{\infty} n^{-2} |f(n)|$.

(c) Exhibit an m-integrable function which is *not* bounded. ∎

To establish the crucial fact that $L^1(m)$ is *complete* (i.e. is a Banach space) for any vector measure $m : \Sigma \longrightarrow X$ we require two further results. The first result is from [29].

Proposition I.6. (a) *Let (Ω, Σ, μ) be a complex measure space and $\lambda_n : \Sigma \longrightarrow \mathbb{C}$ be a sequence of complex measures such that $\lambda_n \ll \mu$ (i.e. $|\lambda_n| \ll |\mu|$), for each $n \in \mathbb{N}$. If $\lim_{n \to \infty} \lambda_n(E)$ exists in $\mathbb{C}$, for each $E \in \Sigma$, then*

$$(15) \qquad \lim_{|\mu|(E) \to 0} \lambda_n(E) = 0, \quad \text{uniformly with respect to } n \in \mathbb{N},$$

that is, for each $\varepsilon > 0$ there is a $\delta > 0$ such that $\sup_n |\lambda_n(E)| < \varepsilon$, for all $E \in \Sigma$ satisfying $|\mu|(E) < \delta$.

(b) *Let $f : \Omega \longrightarrow \mathbb{C}$ be a function and $\{f_n\}_{n=1}^{\infty} \subseteq L^1(\mu)$ satisfy*
(i) *$f_n \to f$ pointwise on Ω, and*
(ii) *$\{\int_E f_n \, d\mu\}_{n=1}^{\infty}$ is Cauchy in $\mathbb{C}$, for each $E \in \Sigma$.*
Then $f \in L^1(\mu)$ and $\|f_n - f\|_1 \longrightarrow 0$ as $n \to \infty$.

Proof. (a) This is the *Vitali–Hahn–Saks theorem*, a classical result from measure theory; see [14; Chapter III, §7], for example.

(b) For each n, define $\lambda_n(E) := \int_E f_n \, d\mu$ for each $E \in \Sigma$. Then $\lambda_n \ll \mu$, for $n \in \mathbb{N}$, and $\lim_{n \to \infty} \lambda_n(E)$ exists in $\mathbb{C}$, for each $E \in \Sigma$; see assumption (ii). By part (a) we conclude that (15) holds. Fix $\varepsilon > 0$. Define

$$E_n := \{w \in \Omega : |f_n(w) - f(w)| \geq \varepsilon\}, \qquad n \in \mathbb{N}.$$

By assumption (i) we have $\lim_n E_n = \emptyset$ and so $|\mu|(E_n) \longrightarrow 0$ by the σ-additivity of $|\mu|$. By (15) there is $m_0 \in \mathbb{N}$ such that

$$\sup_{m \geq m_0} \left(\sup_{n \in \mathbb{N}} |\lambda_n|(E_m) \right) < \varepsilon.$$

So, for $m \geq m_0$, we have

$$\int_{\Omega} |f - f_m| \, d|\mu| \leq \varepsilon \cdot |\mu|(\Omega \backslash E_m) + \liminf_k (\int_{E_m} |f_k - f_m| \, d|\mu|) \leq \varepsilon(2 + \|\mu\|).$$

This inequality implies the desired conclusion. ∎

Let $m : \Sigma \longrightarrow X$ be a vector measure . The Bartle-Dunford-Schwartz theorem guarantees the existence of a finite measure $\mu : \Sigma \longrightarrow [0, \infty)$ such that $m \ll \mu$. A substantial improvement of this result states that a more "special" μ can be selected: this is the following remarkable fact known as *Rybakov's theorem* ; see [8; Chapter IX]. It should be noted that Rybakov's theorem came some 15 years after the Bartle-Dunford-Schwartz theorem.

Theorem I.10. *Let X be a Banach space and $m : \Sigma \longrightarrow X$ be a vector measure. Then there exists $x' \in X'$ with $\|x'\| = 1$ such that $m \ll |\langle m, x' \rangle|$.*

Exercise 12. Let Ω be a compact , topological Hausdorff space, X be a Banach space and $m : Bo(\Omega) \longrightarrow X$ be a vector measure. Then m is called *regular* if for each $E \in Bo(\Omega)$ and $\varepsilon > 0$ there exist a compact set K and an open set U with $K \subseteq E \subseteq U$ and $\|m\|(U \backslash K) < \varepsilon$.

The measure m is called *scalarly regular* if each complex measure $\langle m, x' \rangle$, for $x' \in X'$, is regular.

Show that m is regular if and only if m is scalarly regular.

Hint: Rybakov's theorem may be useful. ∎

We now come to one of the most important properties of $L^1(m)$; the proof is adapted from [36].

Theorem I.11. *Let X be a Banach space and $m : \Sigma \longrightarrow X$ be a vector measure. Then $L^1(m)$ is complete (i.e. it is a Banach space) and the integration map $I_m : L^1(m) \longrightarrow X$ defined by*

$$I_m([f]_m) := \int_\Omega f \, dm, \qquad [f]_m \in L^1(m),$$

is linear and continuous. Moreover, the Σ-simple functions are dense in $L^1(m)$.

Proof. By Theorem I.10 there is $x' \in X'$ with $\|x'\| = 1$ such that $m \ll |\langle m, x' \rangle|$. Let $\{f_n\}_{n=1}^\infty \subseteq L^1(m)$ be a Cauchy sequence. Since

$$\int_\Omega |f_n - f_r| \, d|\langle m, x' \rangle| \leq \sup_{\|z'\| \leq 1} \int_\Omega |f_n - f_r| \, d|\langle m, z' \rangle| = \| [f_n]_m - [f_r]_m \|_1,$$

we see that $\{f_n\}_{n=1}^\infty$ is also Cauchy in $L^1(\langle m, x' \rangle)$. By completeness of $L^1(\langle m, x' \rangle)$ there is $f \in L^1(\langle m, x' \rangle)$ with $f_n \longrightarrow f$ in $L^1(\langle m, x' \rangle)$. Hence, there is a subsequence $\{f_{n_k}\}_{k=1}^\infty$ of $\{f_n\}_{n=1}^\infty$ such that $f_{n_k} \longrightarrow f$ almost everywhere, briefly a.e., with respect to $\langle m, x' \rangle$, [38; Theorem 3.12], and so, also a.e. with respect to m (as $m \ll |\langle m, x' \rangle|$). By redefining f and f_{n_k}, for $k \in \mathbb{N}$, on an m-null set (if necessary) we may assume that $f_{n_k} \longrightarrow f$ pointwise everywhere on Ω, as $k \to \infty$.

Fix $y' \in X'$. Then, for each $E \in \Sigma$ and all $k, \ell \in \mathbb{N}$, we have

$$\left| \int_E f_{n_k} \, d\langle m, y' \rangle - \int_E f_{n_\ell} \, d\langle m, y' \rangle \right| \leq \int_\Omega |f_{n_k} - f_{n_\ell}| \, d|\langle m, y' \rangle|$$

$$= \|y'\| \cdot \int_\Omega |f_{n_k} - f_{n_\ell}| \, d|\langle m, \tfrac{y'}{\|y'\|} \rangle| \leq \|y'\| \cdot \|f_{n_k} - f_{n_\ell}\|_{L^1(m)}.$$

Hence, $\{\int_E f_{n_k} \, d\langle m, y' \rangle\}_{k=1}^\infty$ is Cauchy in $\mathbb{C}$, for each $E \in \Sigma$. So, Proposition I.6(b) (with $\mu = \langle m, y' \rangle$) implies that $f \in L^1(\langle m, y' \rangle)$ and

$$(16) \qquad \lim_{k \to \infty} f_{n_k} = f, \quad \text{in } L^1(\langle m, y' \rangle).$$

Fix $E \in \Sigma$. Then the inequality

$$\left\| \int_E (f_{n_k} - f_{n_\ell}) \, dm \right\| \leq \|f_{n_k} - f_{n_\ell}\|_{L^1(m)}, \qquad k, \ell \in \mathbb{N},$$

shows that $\{\int_E f_{n_k} \, dm\}_{k=1}^\infty$ is Cauchy in X and so, by completeness of X, there is $x_E \in X$ such that $\int_E f_{n_k} \, dm \longrightarrow x_E$ in X, as $k \to \infty$. Then (16) implies that

$$\langle x_E, y' \rangle = \lim_{k \to \infty} \langle \int_E f_{n_k} \, dm, y' \rangle = \lim_{k \to \infty} \int_E f_{n_k} \, d\langle m, y' \rangle = \int_E f \, d\langle m, y' \rangle.$$

This establishes that $f \in L^1(m)$ and $\int_E f \, dm = x_E$, for each $E \in \Sigma$.

Let $\varepsilon > 0$. Then there exists N_ε such that

$$(17) \qquad\qquad \| [f_{n_k}]_m - [f_{n_\ell}]_m \|_1 \leq \varepsilon/4, \qquad k, \ell \geq N_\varepsilon.$$

Fix $E \in \Sigma$. Then, for each $k \geq N_\varepsilon$, we have by (17) that

$$\Big\| \int_E f \, dm - \int_E f_{n_k} \, dm \Big\| = \lim_{\ell \to \infty} \Big\| \int_E f_{n_\ell} \, dm - \int_E f_{n_k} \, dm \Big\| \leq \varepsilon/4.$$

By Proposition I.2 (applied to each vector measure $E \mapsto \int_E (f - f_{n_k}) \, dm$, for $k \in \mathbb{N}$) we see, for all $k \geq N_\varepsilon$, that

$$\| [f]_m - [f_{n_k}]_m \|_1 = \| m_{(f - f_{n_k})} \|(\Omega) \leq 4 \sup_{E \in \Sigma} \Big\| \int_E f \, dm - \int_E f_{n_k} \, dm \Big\| \leq 4 . \varepsilon/4 = \varepsilon.$$

This shows that $[f_{n_k}]_m \longrightarrow [f]_m$ in $L^1(m)$, as $k \to \infty$.

Using the fact that if $\{x_n\}_{n=1}^\infty$ is a Cauchy sequence in a normed space X and $\{x_{n_k}\}_{k=1}^\infty$ is a subsequence with $x_{n_k} \longrightarrow x$ (as $k \to \infty$), for some $x \in X$, then also $x_n \to x$ as $n \to \infty$, it follows that $[f_n]_m \longrightarrow [f]_m$ in $L^1(m)$, as $n \to \infty$. This shows that $L^1(m)$ is complete.

Since $\| \int_\Omega g \, dm \| \leq \|g\|_{L^1(m)}$, for each $g \in L^1(m)$, it is clear that the integration map $I_m : L^1(m) \to X$ is continuous. Linearity of I_m is obvious.

To see that $\mathrm{sim}(\Sigma)$, the space of all Σ-simple functions, is dense in $L^1(m)$ it suffices to consider $[f]_m \in L^1(m)$ with $[f]_m \geq 0$. So, redefine f (if necessary) to be zero on the m-null set $\{w \in \Omega : f(w) \notin [0, \infty)\}$. Choose Σ-simple functions $\{s_n\}_{n=1}^\infty$ such that $0 \leq s_n \uparrow f$ pointwise everywhere on Ω. By the dominated convergence theorem for vector measures (c.f. Theorem I.9) we conclude that

$$\| [s_n]_m - [f]_m \|_{L^1(m)} = \| m_f - m_{s_n} \|(\Omega) \longrightarrow 0, \qquad n \to \infty,$$

and hence, $[s_n]_m \to [f]_m$ in $L^1(m)$, as $n \to \infty$.   ∎

We end this chapter with the vector-valued Riesz representation theorem. Let Ω be a compact , topological Hausdorff space. Then we have seen that the Banach space dual of $(C(\Omega), \| \cdot \|_\infty)$ is the space $M(Bo(\Omega))$ of all *regular* complex measures $\nu : Bo(\Omega) \longrightarrow \mathbb{C}$ equipped with the total variation norm $\|\nu\| = |\nu|(\Omega)$. Hence, it makes sense to talk about the weak-star topology $\sigma(M(Bo(\Omega)), C(\Omega))$ on $M(Bo(\Omega))$.

If X and Y are Banach spaces, then $\mathcal{L}(X, Y)$ denotes the Banach space of all continuous linear operators $T : X \longrightarrow Y$ equipped with the *operator norm*

$$\|T\| := \sup\{\|Tx\|_Y : x \in X, \ \|x\|_X \leq 1\}.$$

In the case when $X = Y$ we denote $\mathcal{L}(X, Y)$ simply by $\mathcal{L}(X)$. The *dual (or adjoint) operator* $T' : Y' \longrightarrow X'$ is defined by

$$\langle Tx, y' \rangle = \langle x, T'y' \rangle, \qquad x \in X, \ y' \in Y';$$

it satisfies $\|T\| = \|T'\|$, [14; Chapter VI]. In particular, $T' \in \mathcal{L}(Y', X')$.

An operator $T \in \mathcal{L}(X, Y)$ is called *weakly compact* if $\{Tx : x \in X, \|x\| \le 1\}$ is a relatively weakly compact subset of Y. The next two results can be found in [14; Chapter VI, §7], for example.

Theorem I.12. *Let Ω be a compact Hausdorff space, X be a Banach space and $T \in \mathcal{L}(C(\Omega), X)$. Then there exists a unique function $F : Bo(\Omega) \longrightarrow X''$ satisfying the following properties.*

(a) $\langle x', F(\cdot) \rangle \in M(Bo(\Omega))$, *for each* $x' \in X'$.

(b) *The mapping* $x' \mapsto \langle x', F(\cdot) \rangle$ *is continuous from X' equipped with its weak-star topology $\sigma(X', X)$ into $M(Bo(\Omega))$ equipped with its weak-star topology $\sigma(M(Bo(\Omega)), C(\Omega))$.*

(c) $\langle Tf, x' \rangle = \int_\Omega f(w)\, d\langle x', F(w) \rangle$, *for* $f \in C(\Omega)$ *and* $x' \in X'$.

(d) $\|T\| = \sup \|\sum_{j=1}^n \alpha_j F(E_j)\|_{X''}$, *where the supremum is taken over all finite collections of disjoint Borel sets $\{E_j\}_{j=1}^n$ in Ω and all finite sets of complex numbers $\alpha_1, \ldots, \alpha_n$ with $|\alpha_j| \le 1$, and $n \in \mathbb{N}$ is arbitrary.*

Conversely, if $F : Bo(\Omega) \longrightarrow X''$ satisfies (a) and (b), then the equation (c) defines a continuous linear operator $T \in \mathcal{L}(C(\Omega), X)$ with norm $\|T\|$ given by (d) and such that $T'x' = \langle x', F(\cdot) \rangle$, for each $x' \in X'$.

The above result (see also [8; p.152]) shows that every $T \in \mathcal{L}(C(\Omega), X)$ can be represented by a *finitely additive* vector measure $F : Bo(\Omega) \longrightarrow X''$ which is σ-additive for the weak-star topology $\sigma(X'', X')$ on X''. Hence, if X is *reflexive* (i.e. $X'' = X$), then $F : Bo(\Omega) \longrightarrow X$ is a genuine *regular* vector measure (c.f. Exercise 12) and (c) becomes

$$(18) \qquad Tf = \int_\Omega f dF, \qquad f \in C(\Omega),$$

after noting each $f \in C(\Omega)$ is bounded and $Bo(\Omega)$-measurable and hence, is F-integrable. To get a similar X-valued representation result without requiring X to be reflexive requires a restriction on T. This is the following result, known as the *Riesz representation theorem for vector measures* ; see also [8; p.153].

Theorem I.13. *Let Ω be a compact Hausdorff space, X be a Banach space and $T \in \mathcal{L}(C(\Omega), X)$ be weakly compact. Then there exists a regular vector measure $m : Bo(\Omega) \longrightarrow X$ (necessarily unique) such that*

(a) $\langle m, x' \rangle \in M(Bo(\Omega))$, *for all* $x' \in X'$,

(b) $Tf = \int_\Omega f dm$, *for each* $f \in C(\Omega)$,

(c) $\|T\| = \|m\|(\Omega)$, *and*

(d) $T'x' = \langle m, x' \rangle$, *for each* $x' \in X'$.

Conversely, if $m : Bo(\Omega) \longrightarrow X$ is a vector measure which satisfies the condition (a), then $T : C(\Omega) \longrightarrow X$ defined by (b) is a weakly compact operator with norm given by (c) and whose dual operator T' is given by (d).

Combining Theorem I.13 with the following result shows that (18) actually holds for *arbitrary* operators $T \in \mathcal{L}(C(\Omega), X)$ in a class of Banach spaces X more general than the reflexive spaces.

Theorem I.14. *Let X be a Banach space which does not have any closed subspace isomorphism to c_0. Let Ω be a compact Hausdorff space. Then every continuous linear operator $T : C(\Omega) \longrightarrow X$ is necessarily weakly compact. In particular, (18) holds for some unique regular vector measure $m : Bo(\Omega) \longrightarrow X$.*

The above result can be found in [8; pp.159-160], for example. Exercise 8(b) shows that every weakly sequentially complete Banach space satisfies the hypothesis of Theorem I.14 (for this class of spaces Theorem I.14 can be found in [14; Chapter VI, §7]). However, there exist Banach spaces which do not contain an isomorphic copy of c_0 and fail to be weakly sequentially complete; see [22; p.73], for example.

Chapter II

Abstract Boolean algebras and Stone spaces

The aim of this chapter is to develop in a systematic way the theory of (abstract) Boolean algebras, as far as is needed later in the text. Far more comprehensive discussions of this topic can be found in [16], [24], [26], [28], [40] and [41], for example. The fundamental result is the Stone representation theorem which states that a Boolean algebra B is isomorphic to the Boolean algebra $Co(\Omega_B)$ of all closed-open subsets of some (essentially unique) totally disconnected, compact Hausdorff space Ω_B. In topological spaces of the type Ω_B, called *Stone spaces* , the sets from $Co(\Omega_B)$ form a base for the topology of Ω_B. Certain completeness properties of B (of an algebraic nature) manifest themselves in certain disconnectedness properties of Ω_B (of a topological kind). Two important σ-algebras which arise are the Baire sets $Ba(\Omega_B)$, which comprise the minimal σ-algebra generated by all closed-open sets, and the Borel sets $Bo(\Omega_B)$, which are generated by all open sets. Examples are given to show that $Ba(\Omega_B) \subseteq Bo(\Omega_B)$ is typically a strict inclusion. The Boolean algebra isomorphism $Q : Co(\Omega_B) \longrightarrow B$ as given by Stone's representation theorem (which is always finitely additive) plays a fundamental role. Moreover, if B is abstractly σ-complete (resp. abstractly complete) , then Q has an extension to a Boolean algebra σ-homomorphism $\overline{Q} : Ba(\Omega_B) \longrightarrow B$ (resp. $\widehat{Q} : Bo(\Omega_B) \longrightarrow B$). Such extension theorems for Q, from the algebra of sets $Co(\Omega_B)$ to the σ-algebras $Ba(\Omega_B)$ and $Bo(\Omega_B)$, will play an important role in subsequent chapters where B will be part of a vector space equipped with a topology and it will become important to decide whether or not the extensions $\overline{Q}$ and $\widehat{Q}$ are σ-additive . In the case when they are σ-additive, it will be possible to apply the methods and techniques of vector measures and integration theory as developed in Chapter I.

We begin with some algebraic preliminaries.

Definition II.1. A *partially ordered set* is a non-empty set A together with a relation $\leq$ satisfying the following properties;

(i) $a \leq a$ for all $a \in A$,

(ii) $a \leq b$ and $b \leq a$ implies $a = b$, and

(iii) $a \leq b$ and $b \leq c$ implies $a \leq c$. ■

Example 7.   (a) $A = \mathbb{R}$ with the usual order $\leq$ of real numbers is a partially ordered set.

(b) Let Ω be any non-empty set and A be the set of all subsets of Ω. For each $E, F \in A$ define $E \leq F$ if $E \subseteq F$. Then $(A, \leq)$ is a partially ordered set. ■

If B is a subset of A, with $(A, \leq)$ partially ordered, then an element $a \in A$ is called an *upper bound* of B if $b \leq a$ for all $b \in B$. An upper bound $a \in A$ of B is said to be a *least upper bound* of B if every upper bound c of B satisfies $a \leq c$. In a similar fashion the terms *lower bound* and *greatest lower bound* of B are defined (if they exist). When it exists, the greatest lower (resp. least upper) bound of B is denoted by $\wedge B$ (resp. $\vee B$).

Definition II.2.   A partially ordered set $(L, \leq)$ is called a *lattice* if every pair $x, y \in L$ has a least upper bound and a greatest lower bound, denoted by $x \vee y$ and $x \wedge y$, respectively.

The lattice L is said to have a *unit* if there exists $1 \in L$ such that $x \leq 1$ for all $x \in L$. If 1^* is another unit, then the conditions $1^* \leq 1$ and $1 \leq 1^*$ together with the properties of a partial order imply that $1^* = 1$, i.e. a unit is *unique* (when it exists).

The lattice L is said to have a *zero* if there exists an element $0 \in L$ (necessarily unique) such that $0 \leq x$ for all $x \in L$.

The lattice L is called *distributive* if $x \vee (y \wedge z) = (x \vee y) \wedge (x \vee z)$ and $x \wedge (y \vee z) = (x \wedge y) \vee (x \wedge z)$, for all $x, y, z \in L$.

The lattice L is *complemented* if it has a unit and zero and if, for every $x \in L$, there is an element $x' \in L$ (called the *complement* of x) such that $x \wedge x' = 0$ and $x \vee x' = 1$. ■

Exercise 13.   (a) Give an example of a partially ordered set which is *not* a lattice.

(b) Give an example of a partially ordered set for which $x \wedge y$ exists for each pair of elements x, y but such that *not* every pair of elements has a least upper bound .

(c) Let $L = \{3, 4, 5, \dots\}$ be equipped with the usual partial order $\leq$ inherited from $\mathbb{R}$. Show that L has a zero element, but no unit. Show that the subset $A = \{4, 6, 8, \dots\}$ does not have an upper bound in L.

(d) Give an example of a lattice which has a unit but no zero element.

(e) Let $L = \{\emptyset, \{1\}, \{2\}, \{3\}, \{1, 2, 3\}\}$ and partially order the elements of L by set inclusion. Show that L is a lattice with zero and unit, but that L is *not* distributive.

(f) Let $(L, \leq)$ be a lattice . For $x, y \in L$ show the following statements are equivalent.

$$\text{(i)} \quad x \leq y \qquad \text{(ii)} \quad x \wedge y = x \qquad \text{(iii)} \quad x \vee y = y.$$ ■

Exercise 14.   Let X be a Banach space and $\mathcal{E} \subseteq \mathcal{L}(X)$ be the family of all (continuous) projection operators on X. Define $E \leq F$ to mean $EF = E = FE$.

(a) Show that $\leq$ is a partial order such that $0 \leq E \leq I$ for all $E \in \mathcal{E}$, where I denotes the identity operator on X.

(b) Show that each $E \in \mathcal{E}$ has a complement, namely $(I - E) \in \mathcal{E}$.

(c) Let $E, F \in \mathcal{E}$ satisfy $EF = FE$. Show that EF is a projection which is the greatest lower bound of E and F in $\mathcal{E}$ and that the range $(EF)X$ of EF is the intersection of the ranges EX and FX.

(d) Let $E, F \in \mathcal{E}$ satisfy $EF = FE$. Show that $E + F - EF$ is a projection which is the least upper bound of E and F in $\mathcal{E}$ and that its range $(E + F - EF)X$ is the linear subspace of X spanned by the set $EX \cup FX$. ∎

Definition II.3. A lattice with zero and unit which is both complemented and distributive is called a *Boolean algebra*, briefly B.a.. ∎

Example 8. (a) Let Ω be a non-empty set and let $\mathcal{B} = 2^\Omega$ be the collection of all subsets of Ω, partially ordered via set inclusion. Then the empty set $\emptyset$ is the zero element, the whole set Ω is the unit, the B.a. complement A' of $A \in \mathcal{B}$ is the set theoretic complement $A^c := \Omega \backslash A$, and the distributive law holds because of the set identity $A \cap (B \cup C) = (A \cap B) \cup (A \cap C)$ and the fact that $A \wedge B = A \cap B$ and $A \vee B = A \cup B$. Hence, $\mathcal{B}$ is a B.a..

(b) Let (Ω, Σ) be any measurable space. Then Σ is a B.a. with respect to the operations of part (a). The same is true if Σ is merely an algebra of sets. ∎

Let $\mathcal{B}$ be a B.a.. Define multiplication and addition in $\mathcal{B}$ by

$$(1) \qquad xy := x \wedge y \quad \text{and} \quad x + y := (x \wedge y') \vee (x' \wedge y), \qquad x, y \in \mathcal{B}.$$

It can be verified that with these operations $\mathcal{B}$ becomes a *Boolean ring* with the unit $1 \in \mathcal{B}$ as its identity; see [14; Chapter I, §12] for the definition of a Boolean ring. On the other hand, if $\mathcal{R}$ is a Boolean ring with identity e and we define a relation $\leq$ in $\mathcal{R}$ by $x \leq y$ if $xy = x$ and complements by $x' = e + x$, then $\mathcal{R}$ becomes a B.a. with

$$x \vee y = x + y - xy \quad \text{and} \quad x \wedge y = xy, \qquad x, y \in \mathcal{R}.$$

Of course, the B.a. unit is the element e. Concerning the B.a. of Example 8(a) we see that the Boolean ring operations are given by intersection for multiplication and symmetric difference for addition, that is, $AB = A \cap B$ and $A + B = A \triangle B := (A \cap B^c) \cup (A^c \cap B)$, for each A, B.

Exercise 15. Let $\mathcal{B}$ be a Boolean ring. Show that $x^2 = x$ and $x + x = 0$, for $x \in \mathcal{B}$. ∎

Exercise 16. [*] Let $\mathcal{B}$ be a Boolean algebra.

(a) Show that $(a')' = a$, for all $a \in \mathcal{B}$, and that $(a \vee b)' = a' \wedge b'$ and $(a \wedge b)' = a' \vee b'$, for all $a, b \in \mathcal{B}$. Deduce that $a \leq b$ if and only if $b' \leq a'$.

(b) Let $\{b_\alpha\}$ be a subset of $\mathcal{B}$ with a least upper bound. For each $b \in \mathcal{B}$, show that the subset $\{b \wedge b_\alpha\}$ also has a least upper bound and that $b \wedge (\vee_\alpha b_\alpha) = \vee_\alpha (b \wedge b_\alpha)$. Show that $\{b'_\alpha\}$ also has a greatest lower bound and that $\wedge_\alpha b'_\alpha = (\vee_\alpha b_\alpha)'$.

(c) Let $\{c_\alpha\}$ be a subset of $\mathcal{B}$ with a greatest lower bound. For each $c \in \mathcal{B}$, show that the subset $\{c \vee c_\alpha\}$ also has a greatest lower bound and that $c \vee (\wedge_\alpha c_\alpha) = \wedge_\alpha (c \vee c_\alpha)$. Show that $\{c'_\alpha\}$ also has a least upper bound and that $\vee_\alpha c'_\alpha = (\wedge_\alpha c_\alpha)'$. ∎

Definition II.4. Let $(\mathcal{B}, \leq)$ be a B.a.. Then $\mathcal{B}$ is called *abstractly complete* (resp. *abstractly σ-complete*) if every subset A of $\mathcal{B}$ (resp. every countable subset A of $\mathcal{B}$) has a greatest lower bound, denoted by $\wedge A$, or equivalently, if every subset A (resp. every countable subset A) has a least upper bound, denoted by $\vee A$. ∎

Exercise 17. (a) Let $\mathcal{B}$ be the collection of all subsets of $\mathbb{N}$ which are finite or have finite complement. Define $A \leq B$ if $A \subseteq B$. Show that $(\mathcal{B}, \leq)$ is a B.a. which is *not* abstractly σ-complete.

(b) Let $\mathcal{B}$ be the collection of all subsets of the interval $[0,1]$ which are countable or have countable complement. Define $A \leq B$ if $A \subseteq B$. Show that $(\mathcal{B}, \leq)$ is a B.a. which is abstractly σ-complete, but *not* abstractly complete. ∎

Definition II.5.   Let $\mathcal{A}$ and $\mathcal{B}$ be B.a.'s and $\Phi : \mathcal{A} \longrightarrow \mathcal{B}$ be a function. Then Φ is called a *homomorphism* if

  (i) $(\Phi(x))' = \Phi(x')$, for $x \in \mathcal{A}$,

  (ii) $\Phi(x \wedge y) = \Phi(x) \wedge \Phi(y)$, for $x, y \in \mathcal{A}$, and

  (iii) $\Phi(x \vee y) = \Phi(x) \vee \Phi(y)$, for $x, y \in \mathcal{A}$.

If, in addition, Φ is injective and satisfies $\Phi(\mathcal{A}) = \mathcal{B}$, then we call Φ an *isomorphism* . ∎

Since $x \wedge 0_{\mathcal{A}} = 0_{\mathcal{A}}$ and $x \vee 1_{\mathcal{A}} = 1_{\mathcal{A}}$, for all $x \in \mathcal{A}$, it follows from (ii) and (iii) that $\Phi(0_{\mathcal{A}}) = 0_{\mathcal{B}}$ and $\Phi(1_{\mathcal{A}}) = 1_{\mathcal{B}}$ whenever Φ is a homomorphism . Moreover, $x \leq y$ implies that $\Phi(x) \leq \Phi(y)$, since $x \leq y$ if and only if $x \wedge y = x$ (c.f. Exercise 13). If Φ is also injective , then $\Phi(x) \leq \Phi(y)$ implies that $x \leq y$. Indeed, $\Phi(x) \leq \Phi(y)$ implies that $\Phi(x) \wedge \Phi(y) = \Phi(x)$. But, by (ii), we also have that $\Phi(x) \wedge \Phi(y) = \Phi(x \wedge y)$ and so $\Phi(x \wedge y) = \Phi(x)$. Then the injectivity of Φ yields $x = x \wedge y$, that is, $x \leq y$.

Exercise 18.[∗]   Let $\mathcal{A}$ and $\mathcal{B}$ be Boolean algebras and $\Phi : \mathcal{A} \longrightarrow \mathcal{B}$ be an *isomorphism* .

(a) Suppose $\{x_\alpha\}$ is a set in $\mathcal{A}$ such that $\vee_\alpha x_\alpha$ exists in $\mathcal{A}$. Show that the least upper bound $\vee_\alpha \Phi(x_\alpha)$ of $\{\Phi(x_\alpha)\}$ exists in $\mathcal{B}$ and $\vee_\alpha \Phi(x_\alpha) = \Phi(\vee_\alpha x_\alpha)$.

(b) Suppose $\{y_\beta\}$ is a set in $\mathcal{A}$ such that $\wedge_\beta y_\beta$ exists in $\mathcal{A}$. Show that the greatest lower bound $\wedge_\beta \Phi(y_\beta)$ of $\{\Phi(y_\beta)\}$ exists in $\mathcal{B}$ and $\wedge_\beta \Phi(y_\beta) = \Phi(\wedge_\beta y_\beta)$.

(c) Deduce that $\mathcal{A}$ is abstractly complete (resp. σ-complete) if and only if $\mathcal{B}$ is abstractly complete (resp. σ-complete). ∎

We now turn our attention to some topological considerations.

Suppose that Ω is a compact Hausdorff space. Denote by $Co(\Omega)$ the collection of all subsets of Ω which are simultaneously open and closed (such sets are called *clopen*) . Note that always $\emptyset \in Co(\Omega)$ and $\Omega \in Co(\Omega)$. It is easy to exhibit examples where these are the only elements of $Co(\Omega)$. The collection of sets $Co(\Omega)$ turns out to be a B.a. with respect to the operations given in Example 8; this follows from elementary properties of open and closed sets in a topological space. This particular B.a. $Co(\Omega)$ will play a fundamental role in the sequel. We point out that $Co(\Omega)$ is also an *algebra of sets* in the sense of measure theory; see Chapter I.

Definition II.6.   A compact topological Hausdorff space Ω is called *totally disconnected* if $Co(\Omega)$ forms a *base* for the topology in Ω. That is, every open set in Ω is the union of some subcollection from $Co(\Omega)$. ∎

Exercise 19. (a) Let $\Omega = \{0\} \cup \{\frac{1}{n} : n \in \mathbb{N}\}$ be equipped with the relative topology from $\mathbb{R}$. Describe all the clopen subsets of Ω. Show that Ω is totally disconnected.

(b) Let $\Omega = [0,1]$ with its usual topology. Using the fact that every open set in Ω is a countable union of pairwise disjoint, open intervals from Ω describe all the clopen subsets of Ω. Show that Ω is *not* totally disconnected . ∎

Given a topological Hausdorff space Ω, let $\text{sim}(Co(\Omega))$ denote the vector space of all functions $f : \Omega \longrightarrow \mathbb{C}$ which are a finite linear combination of functions of the form χ_E, where $E \in Co(\Omega)$. Each $f \in \text{sim}(Co(\Omega))$ has a unique expression

$$(2) \qquad f = \sum_{j=1}^{n} \alpha_j \chi_{E_j},$$

called its *standard representation* , where the complex numbers $\{\alpha_j\}_{j=1}^n$ are all *distinct* and $\{E_j\}_{j=1}^n \subseteq Co(\Omega)$ is a finite family of non-empty, pairwise disjoint sets with $\cup_{j=1}^n E_j = \Omega$.

Lemma II.1. *Let Ω be a compact Hausdorff space.*

(a) *The vector space $\text{sim}(Co(\Omega))$ is also a unital algebra of functions, meaning that $fg \in \text{sim}(Co(\Omega))$ whenever $f, g \in \text{sim}(Co(\Omega))$. The unit is the constant function $\mathbb{1} : w \mapsto 1$, for $w \in \Omega$.*

(b) *The algebra $\text{sim}(Co(\Omega))$ is conjugate closed, that is, $\overline{f} \in \text{sim}(Co(\Omega))$ whenever $f \in \text{sim}(Co(\Omega))$, where $\overline{f}$ denotes the function $w \mapsto \overline{f(w)}$, for $w \in \Omega$.*

(c) *If $f \in \text{sim}(Co(\Omega))$ has standard representation given by (2), then $\frac{1}{f} \in \text{sim}(Co(\Omega))$ if and only if $0 \notin \{\alpha_j\}_{j=1}^n$. Then $\frac{1}{f} = \sum_{j=1}^n (\frac{1}{\alpha_j}) \chi_{E_j}$ is the standard representation of $\frac{1}{f}$.*

(d) *If (2) is the standard representation of $f \in \text{sim}(Co(\Omega))$, then*

$$\|f\|_\infty = \max\{|\alpha_j| : 1 \le j \le n\}.$$

Exercise 20. Prove Lemma II.1. ∎

Let Λ be a set and $\mathcal{F}$ be a family of $\mathbb{C}$-valued functions on Λ. Then we say that $\mathcal{F}$ *distinguishes the points of* Λ if, for every pair of distinct points $\lambda_1, \lambda_2 \in \Lambda$, there exists $\phi \in \mathcal{F}$ such that $\phi(\lambda_1) \ne \phi(\lambda_2)$.

Example 9. Let Ω be a compact Hausdorff space. A result from topology (called *Urysohn's theorem*) states for such spaces Ω that if A, B are disjoint, closed sets in Ω, then there exists a continuous function $f : \Omega \longrightarrow \mathbb{C}$ with $f(\Omega) \subseteq [0, 1]$ such that $f(A) = \{0\}$ and $f(B) = \{1\}$; see [14; Chapter I, §5], for example. Since singleton sets $\{w\}$, for $w \in \Omega$, are closed sets it follows that $C(\Omega)$ distinguishes the points of Ω. ∎

The following result shows that for certain topological spaces Ω the algebra $C(\Omega)$ in Example 9 can be replaced by a smaller subalgebra.

Lemma II.2. *Let Ω be a compact Hausdorff space which is totally disconnected . Then $\text{sim}(Co(\Omega))$ distinguishes the points of Ω.*

Proof. Let w_1 and w_2 be distinct points of Ω. Since Ω is Hausdorff, there exist disjoint open sets U_1 and U_2 in Ω with $w_1 \in U_1$ and $w_2 \in U_2$. But, $Co(\Omega)$ forms a base for the topology of Ω, and so there is a set $W \in Co(\Omega)$ with $w_1 \in W$ and $W \subseteq U_1$. Then $\chi_W \in \text{sim}(Co(\Omega))$ satisfies $\chi_W(w_1) = 1$, but $\chi_W(w_2) = 0$. ∎

We now recall a classical result concerning function algebras, namely the *Stone-Weierstrass theorem* ; see [14; Chapter IV, §6], for example.

Theorem II.1. *Let Ω be a compact Hausdorff space and $\mathcal{A}$ be a norm closed subalgebra of $C(\Omega)$ satisfying both*

(i) *$\mathbb{1} \in \mathcal{A}$, and*

(ii) $\overline{f} \in \mathcal{A}$ whenever $f \in \mathcal{A}$.

Then $\mathcal{A} = C(\Omega)$ if and only if $\mathcal{A}$ distinguishes the points of Ω.

As an immediate consequence we have the following useful result.

Proposition II.1. *Let Ω be a compact Hausdorff space which is totally disconnected. Then $\text{sim}(Co(\Omega))$ is dense in $C(\Omega)$.*

Proof. By Lemma II.1 we know that $\mathcal{F} := \text{sim}(Co(\Omega))$ is a conjugate closed, unital subalgebra of $C(\Omega)$. Then $\mathcal{A} := \overline{\mathcal{F}}$ (the bar denotes closure) is a norm closed subalgebra of $C(\Omega)$ with properties (i) and (ii) of Theorem II.1. Since $\mathcal{F}$ distinguishes the points of Ω (c.f. Lemma II.2) so does $\mathcal{A}$. Then the Stone-Weierstrass theorem implies that $\mathcal{A} = C(\Omega)$.   ■

We can now formulate the fundamental representation theorem for Boolean algebras, the so called *Stone representation theorem*, due to M.H. Stone; see [14; Chapter I, §13].

Theorem II.2. *Let $\mathcal{B}$ be a B.a.. Then there exists a totally disconnected, compact Hausdorff space $\Omega_{\mathcal{B}}$, unique up to topological homeomorphism, such that $\mathcal{B}$ is isomorphic to the B.a. $Co(\Omega_{\mathcal{B}})$.*

Remark. (a) The uniqueness of $\Omega_{\mathcal{B}}$, up to homeomorphism, follows from the fact that if Ω_1 and Ω_2 are compact Hausdorff spaces such that $C(\Omega_1)$ and $C(\Omega_2)$ are algebraically isomorphic, then Ω_1 and Ω_2 are topologically homeomorphic, [14; p.279].

(b) The totally disconnected, compact Hausdorff space $\Omega_{\mathcal{B}}$ as given by Theorem II.2 is called the *Stone space* of $\mathcal{B}$. Moreover, the B.a. isomorphism $Q : Co(\Omega_{\mathcal{B}}) \longrightarrow \mathcal{B}$ given by Theorem II.2 is called the *Stone map*.

(c) The Stone map Q is *finitely additive*. That is, if we turn $\mathcal{B}$ into a Boolean ring with multiplication and addition as given by (1), then

$$(3) \qquad\qquad Q(\cup_{j=1}^{n} A_j) = Q(A_1) + \ldots + Q(A_n)$$

whenever $\{A_j\}_{j=1}^{n} \subseteq Co(\Omega_{\mathcal{B}})$ is a finite collection of *pairwise disjoint* sets. We establish (3) when $n = 2$; the general case follows by induction. So, let $A_1, A_2 \in Co(\Omega_{\mathcal{B}})$ satisfy $A_1 \cap A_2 = \emptyset$. Then it follows from (1) and the fact that Q is a homomorphism (c.f. Definition II.5), that

$$Q(A_1) + Q(A_2) := [Q(A_1) \wedge Q(A_2)'] \vee [Q(A_1)' \wedge Q(A_2)] = Q((A_1 \wedge A_2') \vee (A_1' \wedge A_2)).$$

But, by definition of the operations $\wedge, \vee$ and complementation in $Co(\Omega_{\mathcal{B}})$ we have that

$$(A_1 \wedge A_2') \vee (A_1' \wedge A_2) = (A_1 \cap A_2^c) \cup (A_1^c \cap A_2) = A_1 \cup A_2,$$

(where the last equality uses the disjointness of A_1 and A_2) which implies that $A_1 \cap A_2^c = A_1$ and $A_1^c \cap A_2 = A_2$. Accordingly, $Q(A_1) + Q(A_2) = Q(A_1 \cup A_2)$ as required.

(d) Let Ω be a *given* totally disconnected, compact Hausdorff space. Define $\mathcal{B}$ to be the B.a. $Co(\Omega)$. Then $\Omega_{\mathcal{B}} = \Omega$; see [40; p.25].   ■

Exercise 21. Let $\mathcal{B}$ be a B.a. and consider $\mathcal{B}$ also as a Boolean ring with respect to addition and multiplication as given by (1). Show that $x + y = x \vee y$ whenever $x, y \in \mathcal{B}$ are *disjoint*, meaning that $x \wedge y = 0$.

Hint: The identities $x = x \wedge 1 = x \wedge (y' \vee y)$ and $y = 1 \wedge y = (x' \vee x) \wedge y$ may be useful. ∎

For a given B.a. $\mathcal{B}$ it is not always easy to identify its Stone space $\Omega_{\mathcal{B}}$. We proceed to give two examples of a quite general nature; further examples can be found in [16], [26], [28], [40]. But first we require some further concepts from topology.

Recall that the *discrete topology* on a non-empty set Λ is simply the collection $\tau := 2^{\Lambda}$ consisting of all subsets of Λ.

Let (Λ, τ) be any *locally compact* Hausdorff space, meaning that for each $\lambda \in \Lambda$ there exists an open set $U_{\lambda} \in \tau$ containing λ such that its closure $\overline{U_{\lambda}}$ is compact. Let w be any element *not* in Λ and define $\Lambda_{\infty} = \Lambda \cup \{w\}$. Then the collection τ_{∞} consisting of all subsets of Λ_{∞} which are either open sets in Λ (i.e. belong to τ) or the complement (in Λ_{∞}) of compact subsets of Λ, form a topology in Λ_{∞}. The topological space $(\Lambda_{\infty}, \tau_{\infty})$ is called the *one-point compactification* of (Λ, τ).

Exercise 22. (a) Let Λ be a non-empty set equipped with the discrete topology . Show that Λ is both locally compact and Hausdorff.

(b) Let (Λ, τ) be any locally compact Hausdorff space and $(\Lambda_{\infty}, \tau_{\infty})$ be its one-point compactification . Show that τ_{∞} is indeed a *topology* , that is,

(i) $\emptyset, \Lambda_{\infty} \in \tau_{\infty}$,

(ii) $U \cap V \in \tau_{\infty}$ whenever $U, V \in \tau_{\infty}$, and

(iii) $\cup_{\alpha} V_{\alpha} \in \tau_{\infty}$ for every family of open sets $\{V_{\alpha}\} \subseteq \tau_{\infty}$.

Equip $\Omega := \Lambda_{\infty} \backslash \{w\}$ with the *relative topology* from Λ_{∞} (of course, $\Omega = \Lambda$ as a set), meaning that a set $U \subseteq \Omega$ is defined to be open if it is of the form $U = \Omega \cap V$ for some $V \in \tau_{\infty}$. Show that the identity function from (Λ, τ) onto Ω is a topological homeomorphism. Moreover, verify that $(\Lambda_{\infty}, \tau_{\infty})$ is indeed a compact Hausdorff space. Show that Λ is dense in the space $(\Lambda_{\infty}, \tau_{\infty})$. ∎

A topological space (Λ, τ) is called *completely regular* if, given any closed set $B \subseteq \Lambda$ and a point $\lambda \in B^{c}$, there exists a continuous function $f : \Lambda \longrightarrow \mathbb{R}$ such that $f(B) = \{0\}$ and $f(\lambda) = 1$. Every compact Hausdorff space is completely regular; this follows easily from Urysohn's theorem (c.f. Example 9).

A *compactification* of a topological space Λ is a compact topological space K together with a continuous, injective map $\Phi : \Lambda \longrightarrow K$ such that $\Phi(\Lambda)$ is *dense* in K, that is, $\overline{\Phi(\Lambda)} = K$. The one-point compactification of a locally compact, non-compact space is an example of a compactification; see Exercise 22(b). A more interesting compactification is that due to M.H. Stone and E. Čech ; see [24; Chapter 6, §5] or [41], for example.

Theorem II.3. *Let Λ be a completely regular Hausdorff space. Then Λ has a compactification $\beta(\Lambda)$ with the property that every continuous map from Λ into any compact Hausdorff space Ω has a continuous extension from $\beta(\Lambda)$ into Ω.*

Furthermore, $\beta(\Lambda)$ is unique, in the following sense: if a compactification K of Λ also has the property that every continuous map from Λ into any compact Hausdorff space Ω has a continuous extension from K into Ω, then there exists a homeomorphism of $\beta(\Lambda)$ onto K that leaves Λ pointwise fixed.

The compact Hausdorff space $\beta(\Lambda)$ is referred to as the *Stone-Čech compactification* of Λ. If Λ is already compact, then $\beta(\Lambda)$ is homeomorphic to Λ.

Now to the two examples alluded to above.

Example 10. (a) Let Λ be a non-empty set and $\mathcal{B} := 2^\Lambda$ be the B.a. of all subsets of Λ (c.f. Example 8). If we equip Λ with the discrete topology , then it is routine to verify that Λ is a completely regular Hausdorff space. It turns out that $\Omega_\mathcal{B} = \beta(\Lambda)$; see [40; p.26].

(b) Let Λ be an infinite set and $\mathcal{B}$ denote the collection of all subsets of Λ which are either finite or the complement of a finite set. Then $\mathcal{B}$ is a B.a. (c.f. Exercise 17). If we equip Λ with its discrete topology, then it is a locally compact Hausdorff space (c.f. Exercise 22(a)). It turns out that $\Omega_\mathcal{B} = \Lambda_\infty$ is the one-point compactification of Λ; see [40; p.26]. ∎

The B.a.'s of projections that we will be dealing with in later chapters will often be complete, or at least σ-complete; see Definition II.4. For such B.a.'s more can be said about their Stone spaces than in general; see Proposition II.4 below. So, we now concentrate on properties of this more specialized class of B.a.'s, for which some topological notions are required.

Definition II.7. Let Ω be a compact Hausdorff space.

(i) The space Ω is called *basically disconnected in the restricted sense* if the closure of every *countable* union of sets from $Co(\Omega)$ is an open set.

(ii) The space Ω is called *extremely disconnected* if the closure of every open set is again an open set. ∎

Example 11. (a) Let $\Omega = \{0\} \cup \{\frac{1}{n} : n \in \mathbb{N}\}$ be the totally disconnected , compact Hausdorff space of Exercise 19(a). For each $n \in \mathbb{N}$, let $U_n = \{\frac{1}{2n}\}$ in which case $U_n \in Co(\Omega)$. Then the closure $\overline{\cup_{n=1}^\infty U_n}$ equals $\{0\} \cup \{\frac{1}{2n} : n \in \mathbb{N}\}$ which is *not* an open set in Ω. Indeed, its complement is the set $E = \{\frac{1}{(2n-1)} : n \in \mathbb{N}\}$. Since $0 \in \overline{E} \backslash E$ we see that E is not closed and so its complement $\overline{\cup_{n=1}^\infty U_n}$ is not open. Accordingly, Ω is *not* basically disconnected in the restricted sense .

(b) Every extremely disconnected space is obviously basically disconnected in the restricted sense. ∎

In relation to Example 11(b) the following result is useful in producing examples of *compact* extremely disconnected spaces.

Proposition II.2. *Let Λ be a completely regular Hausdorff space.*

(a) *An isolated point of Λ is also isolated in $\beta(\Lambda)$.*

(b) *Λ is an open set in $\beta(\Lambda)$ if and only if Λ is locally compact .*

(c) *If $A \in Co(\Lambda)$, then the closure (in $\beta(\Lambda)$) of A and $\Lambda \backslash A$ are open sets in $\beta(\Lambda)$ which are complements of one another (in $\beta(\Lambda)$). In particular, they belong to $Co(\beta(\Lambda))$, are pairwise disjoint and have union equal to $\beta(\Lambda)$.*

(d) *Λ is extremely disconnected (using the same formulation as in Definition II.7(ii)) if and only if $\beta(\Lambda)$ is extremely disconnected.*

(e) *If Λ is compact , then Λ is extremely disconnected if and only if $\Lambda = \beta(Y)$ for every dense subspace Y of Λ.*

For (a), (b) and (c) we refer to [24; p.90] and for (d) and (e) we refer to [24; p.96].

Example 12. Let Λ be any non-empty set equipped with its discrete topology. Then Λ is clearly extremely disconnected and completely regular. So, Proposition II.2(d) shows that $\beta(\Lambda)$ is compact and extremely disconnected. ∎

The question arises of whether there is an analogue of Proposition II.2(d) for basically disconnected spaces. We proceed to show that this is the case; see Proposition II.3 below.

Let Λ be a topological Hausdorff space and $C_{\mathbb{R}}(\Lambda)$ denote the vector space of all continuous functions $f : \Lambda \longrightarrow \mathbb{R}$. Any set of the form $f^{-1}((0, \infty))$ or $f^{-1}((-\infty, 0))$ is called a *co-zero set*. In the following definition the topological space is *not* assumed to be compact (c.f. Definition II.6 and Definition II.7).

Definition II.8. (i) A topological Hausdorff space is called *basically disconnected* if the closure of every co-zero set is an open set.

(ii) A completely regular Hausdorff space is called *totally disconnected* if its only connected subsets are those containing a single point. ∎

Remark. For a *compact* space Λ, Definition II.8(ii) is equivalent to the requirement that $Co(\Lambda)$ forms a base for the topology in Λ, [24; p.247]. Accordingly, for *compact* Hausdorff spaces, Definition II.8(ii) agrees with Definition II.6. ∎

Proposition II.3. (a) *Let Λ be a completely regular Hausdorff space. The space Λ is basically disconnected if and only if $\beta(\Lambda)$ is basically disconnected.*

(b) *Suppose that Λ is also compact. Then Λ is basically disconnected, if and only if, Λ is both totally disconnected and basically disconnected in the restricted sense.*

(c) *If Λ is the Stone space of some B.a., then Λ is basically disconnected if and only if it is basically disconnected in the restricted sense.*

Proof. (a) See [24; p.96], for example.

(b) Suppose that Λ is basically disconnected. Let $\{U_n\}_{n=1}^{\infty}$ be a sequence of clopen sets in Λ, in which case $U_n = \{\lambda \in \Lambda : \chi_{U_n}(\lambda) > 0\}$ for each $n \in \mathbb{N}$. By [24; 3N.4, p.52] the closure of $\cup_{n=1}^{\infty} U_n$ is again open. This shows that Λ is basically disconnected in the restricted sense. The Remark after Definition II.8 and [24; 4K.8, p.63] show that Λ is totally disconnected

Now assume that Λ is totally disconnected and basically disconnected in the restricted sense. Let $A = \{\lambda \in \Lambda : f(\lambda) > 0\}$ be a co-zero set, where $f \in C_{\mathbb{R}}(\Lambda)$. Since A is open and $Co(\Lambda)$ forms a base for the topology in Λ we have $A = \cup_{\alpha} A_{\alpha}$ for some sets $A_{\alpha} \in Co(\Lambda)$. Since the closed (hence, *compact*) sets $A^{(n)} := f^{-1}([n, \infty))$, for $n \in \mathbb{N}$, satisfy $A^{(n)} \subseteq \cup_{\alpha} A_{\alpha}$ there is a finite set of indices $F(n)$ such that $A^{(n)} \subseteq \cup_{\alpha \in F(n)} A_{\alpha}$. Then $A := \cup_{n=1}^{\infty} (\cup_{\alpha \in F(n)} A_{\alpha})$ is a *countable* union of sets from $Co(\Lambda)$ and so $\overline{A}$ is open (as Λ is basically disconnected in the restricted sense). This shows that Λ is basically disconnected.

(c) This follows from part (b) and the fact that the Stone space of a B.a. is always a totally disconnected, compact Hausdorff space. ∎

The following result, which will play a crucial role later, makes the connection between certain completeness properties of a B.a. with certain "disconnectedness" properties of its Stone space. We refer to [28; Theorem 7.21] or [41; p.47 & p.69], for example.

Proposition II.4. *Let $\mathcal{B}$ be a B.a.. Then $\mathcal{B}$ is abstractly complete (resp. abstractly σ-complete) if and only if its Stone space $\Omega_{\mathcal{B}}$ is extremely (resp. basically) disconnected.*

It is clear for compact Hausdorff spaces that we have:

Extremely disconnected $\Longrightarrow$ Basically disconnected $\Longrightarrow$ Totally disconnected .

Indeed, the first implication is obvious and the second implication follows from Proposition II.3(b). Example 11(a) shows that the second implication is not an equivalence. To see that the first implication also fails to be an equivalence, let $\mathcal{B}$ be the B.a. of Exercise 17(b), in which case $\mathcal{B}$ is abstractly σ-complete, but not abstractly complete. Then by Proposition II.4 the Stone space $\Omega_{\mathcal{B}}$ of $\mathcal{B}$ is a compact , basically disconnected space which is not extremely disconnected .

Exercise 23. Let Λ be an *infinite* set equipped with its discrete topology τ and let $(\Lambda_\infty, \tau_\infty)$ be the one-point compactification of Λ.

(a) Show $U \subseteq \Lambda_\infty$ is clopen if and only if $U = F$ or $U = \Lambda_\infty \backslash F$ for some finite set $F \subset \Lambda$.

(b) Show that $(\Lambda_\infty, \tau_\infty)$ is totally disconnected .

(c) Show that $(\Lambda_\infty, \tau_\infty)$ is *not* basically disconnected .

(d) Using part (c), deduce that the B.a. $\mathcal{B}$ of Example 10(b) cannot be abstractly σ-complete .

Note: If we let $\Lambda = \{\frac{1}{n} : n \in \mathbb{N}\}$, then Λ_∞ can be identified with the space Ω of Example 11(a). Hence, the phenomenon in Example 11(a) is a special case of part (c) above. ■

Another important class of sets (for our purposes) will be the following one.

Definition II.9. Let Ω be a compact Hausdorff space. Then the σ-algebra of *Baire sets* , denoted by $Ba(\Omega)$, is defined to be the smallest σ-algebra with respect to which each function $f \in C(\Omega)$ is measurable. ■

Clearly $Ba(\Omega) \subseteq Bo(\Omega)$. Moreover, being σ-algebras, both $Ba(\Omega)$ and $Bo(\Omega)$ are abstractly σ-complete as B.a. 's (with respect to the operations given in Example 8(b)), since the supremum (resp. infimum) of each countable subfamily is given by its union (resp. intersection). Also, the algebra of clopen sets satisfies $Co(\Omega) \subseteq Ba(\Omega)$, since $\text{sim}(Co(\Omega)) \subseteq C(\Omega)$. Within the class of Stone spaces the relationship between $Co(\Omega)$ and $Ba(\Omega)$ can be precisely described. Given any family $\mathcal{H}$ of subsets of Ω we let $\sigma(\mathcal{H})$ denote the σ-algebra generated by $\mathcal{H}$.

Proposition II.5. *Let Ω be a compact , totally disconnected Hausdorff space. Then $Ba(\Omega) = \sigma(Co(\Omega))$.*

Proof. Since $Ba(\Omega)$ is a σ-algebra and $Co(\Omega) \subseteq Ba(\Omega)$ it is clear that $\sigma(Co(\Omega)) \subseteq Ba(\Omega)$.

Let $f \in C(\Omega)$. Then there exists a sequence of functions $\{s_n\}_{n=1}^\infty \subseteq \text{sim}(Co(\Omega))$ satisfying $\|s_n - f\|_\infty \longrightarrow 0$, as $n \to \infty$; see Proposition II.1. In particular, $s_n \to f$ pointwise on Ω. Since each function s_n is $\sigma(Co(\Omega))$-measurable, for $n \in \mathbb{N}$, the pointwise limit f is also $\sigma(Co(\Omega))$-measurable. Since $f \in C(\Omega)$ is arbitrary, it follows from Definition II.9 that $Ba(\Omega) \subseteq \sigma(Co(\Omega))$. ■

The *algebra of sets* $Co(\Omega)$, even if $Co(\Omega)$ is abstractly σ-complete or complete as a B.a. , is typically *not* a σ-algebra of sets. The reason is that if $\{E_n\}_{n=1}^\infty \subseteq Co(\Omega)$, then the elements

$\vee_n E_n$ and $\wedge_n E_n$ formed in the B.a. $Co(\Omega)$ are generally *not* given by $\cup_{n=1}^{\infty} E_n$ and $\cap_{n=1}^{\infty} E_n$, respectively. Rather, $\vee_n E_n = \overline{\cup_{n=1}^{\infty} E_n}$ and $\wedge_n E_n$ is the *interior* of $\cap_{n=1}^{\infty} E_n$, denoted by $(\cap_{n=1}^{\infty} E_n)^\circ$. The following example illustrates the point.

Example 13. Let $\mathbb{N}$ have the discrete topology in which case it is clearly extremely disconnected . By Proposition II.2(d) the space $\Omega = \beta(\mathbb{N})$ is compact and extremely disconnected. Since $\Omega_B = \Omega$ (by Remark (d) after Theorem II.2), where $\mathcal{B} := Co(\Omega)$, it follows from Proposition II.4 that the B.a. $Co(\Omega)$ is *abstractly complete*. Proposition II.2(a) shows that each singleton set $\{n\}$, for $n \in \mathbb{N}$, is a clopen set in Ω. Accordingly, $\mathbb{N} = \cup_{n=1}^{\infty}\{n\}$ is certainly a Baire set in Ω. If $Co(\Omega)$ was a σ-algebra , then we would have $\mathbb{N} \in Co(\Omega)$ and so $\mathbb{N} = \overline{\mathbb{N}}$ (the closure taken in $\beta(\mathbb{N})$). But, $\overline{\mathbb{N}} = \Omega$ (as Λ is always dense in $\beta(\Lambda)$) and we have a contradiction. So, $Co(\Omega)$ is *not* a σ-algebra and the inclusion $Co(\Omega) \subseteq Ba(\Omega)$ is strict. ∎

It is always the case that $Ba(\Omega) \subseteq Bo(\Omega)$. Is it likely that this is an equality, with perhaps certain restrictions on Ω?

Exercise 24. Let $\Omega = \{0\} \cup \{\frac{1}{n} : n \in \mathbb{N}\}$ be the totally disconnected , compact Hausdorff space of Exercise 19(a). Using Proposition II.5 show that $Ba(\Omega) = Bo(\Omega)$. ∎

The underlying reason for equality in Exercise 24 is the general fact that $Ba(\Omega) = Bo(\Omega)$ whenever Ω is a compact metric space, [37; pp.302-303], or if Ω has a countable base for its topology , [26; p.100]. Unfortunately, Stone spaces are rarely metrizable. Let us show that $Ba(\Omega) \subseteq Bo(\Omega)$ is *strict* for the example of Exercise 23. For this we will require the following useful fact; see [26; p.99], for example.

Lemma II.3. *Let Ω be a totally disconnected , compact Hausdorff space. Then every open Baire set in Ω is the countable union of sets from $Co(\Omega)$.*

Example 14. Let Λ be an *uncountable* set equipped with its discrete topology τ. Let $\Omega := \Lambda_\infty$ be the one-point compactification of Λ, where $\Lambda_\infty = \Lambda \cup \{w\}$ with $w \notin \Lambda$. Then Ω is compact and totally disconnected ; see Exercise 23.

The first claim is that $Bo(\Omega)$ consists of *all* subsets of Ω, that is, $Bo(\Omega) = 2^\Omega$. Indeed, since $\Lambda \in \tau \subseteq \tau_\infty$, its complement $\{w\} = \Omega \backslash \Lambda$ is a closed set in Ω. In particular, $\{w\} \in Bo(\Omega)$. Moreover, for each $\lambda \in \Lambda$ the singleton set $\{\lambda\} \in Bo(\Omega)$, since $\{\lambda\} \in \tau \subseteq \tau_\infty$. Now let $U \subseteq \Omega$ be arbitrary. If $w \notin U$ then U is open in Ω, since $U \in \tau \subseteq \tau_\infty$. Otherwise, $U = \{w\} \cup V$ where $V = U\backslash\{w\}$ is τ-open in Λ and so τ_∞-open in Ω. That is, $V \in Bo(\Omega)$. Since also $\{w\} \in Bo(\Omega)$ it follows that $U \in Bo(\Omega)$. So, $Bo(\Omega) = 2^\Omega$.

We show the compact set $\{w\} \notin Ba(\Omega)$. Now, Λ is open in Ω. If it were a Baire set , then Lemma II.3 would simply that $\Lambda = \cup_{n=1}^{\infty} E_n$ for some sequence of sets $\{E_n\}_{n=1}^{\infty} \subseteq Co(\Omega)$. In particular, $w \notin E_n$ for every $n \in \mathbb{N}$. So, by Exercise 23(a) each set $E_n \subseteq \Lambda$ is a finite set. Accordingly, Λ is countable which is a contradiction. Then Λ, and hence also $\{w\}$, is *not* a Baire set. ∎

In the previous example the space Ω is totally disconnected , but not basically disconnected: can the same phenomenon occur if Ω is basically or extremely disconnected? Example 15 below shows it can indeed occur in such spaces. First we require a further topological fact.

A regular topological space Λ is called a *Lindelöf space* if every open cover of Λ has a *countable* subcover.

Lemma II.4.   (a) *Let Λ be a compact Hausdorff space and $Y \subseteq \Lambda$ be equipped with the relative topology. If there exists a countable family $\{F_n\}_{n=1}^{\infty}$ of closed subsets of Λ with the property that for every pair of points u, v with $u \in Y$ and $v \in \Lambda \backslash Y$ there is some $n \in \mathbb{N}$ such that $u \in F_n$ and $v \notin F_n$, then Y is a Lindelöf space.*

(b) *Let Ω be a totally disconnected , compact Hausdorff space. Then every open Baire subset of Ω is Lindelöf.*

Proof.   Part (a) can be found in [17; p.250, Ex.3.8F] (be careful of the misprint).

(b) Let Y be an open Baire set in Ω. By Lemma II.3 there is a sequence $\{F_n\}_{n=1}^{\infty} \subseteq Co(\Omega)$ with $Y = \cup_{n=1}^{\infty} F_n$. Let $u \in Y$ and $v \in \Omega \backslash Y$. Then $u \in F_n$ for some n, and certainly $v \notin F_n$ as $v \notin Y$. Part (a) then implies that Ω is a Lindelöf space.   ∎

Example 15.   We give an example of an extremely disconnected, compact Hausdorff space Ω for which the inclusion $Ba(\Omega) \subseteq Bo(\Omega)$ is strict. Let Λ be an *uncountable* set equipped with its discrete topology . By Exercise 22(a) the space Λ is locally compact and Hausdorff, and it is clearly completely regular . Since Λ is obviously extremely disconnected, so is $\Omega := \beta(\Lambda)$ by Proposition II.2(d). Moreover, Proposition II.2(b) implies that Λ is an open subset of Ω. Since each point of Λ is isolated (in Λ), it follows from Proposition II.2(a) that the family of sets $\{\{\lambda\} : \lambda \in \Lambda\}$ is contained in $Co(\Omega)$ and obviously forms an open cover of Λ in Ω. Since this particular open cover of the *open* set $\Lambda \subseteq \Omega$ clearly has no countable subcover, it follows from Lemma II.4(b) that Λ cannot be a Baire set in Ω. But, Λ being an open set, it is immediate that $\Lambda \in Bo(\Omega)$.   ∎

The following exercise provides an example of a basically disconnected , compact Hausdorff space Ω which is not extremely disconnected and for which the inclusion $Ba(\Omega) \subseteq Bo(\Omega)$ is strict.

Exercise 25.   Let Y be an *uncountable* set equipped with its discrete topology τ. Let w be a point not in Y and let $\Lambda = Y \cup \{w\}$. Define a collection of sets ρ in Λ to consist of all sets from τ together with all sets of the form $\Lambda \backslash C$, where C is a *countable* subset of Y.

(a) Verify that ρ is a topology on Λ.

(b) Show that the topology ρ is Hausdorff.

(c) Show that (Λ, ρ) is completely regular .

(d) Verify that (Λ, ρ) is *not* locally compact.

The space (Λ, ρ) is basically disconnected, but not extremely disconnected; see [24; p.64], for example. Let $\Omega := \beta(\Lambda)$.

(e) Show that Ω is a compact , basically disconnected Hausdorff space.

(f) Show that Λ is *not* an open subset of Ω (*Hint:* Use part (d) and Proposition II.2(b)).

(g) Show that each point of Y is an isolated point of (Λ, ρ) and hence, is also an isolated point of Ω. Deduce that Y is an open subset of Ω for which there exists an open cover with no countable subcover.

(h) Using (g), prove that $Y \in Bo(\Omega)$ but $Y \notin Ba(\Omega)$.   ∎

The Stone representation theorem makes the precise connection between B.a.'s and totally disconnected, compact Hausdorff spaces. Namely, a B.a. $\mathcal{B}$ is isomorphic to $Co(\Omega_B)$, where Ω_B is the Stone space of $\mathcal{B}$. It was observed that $Co(\Omega_B)$ is always an algebra of sets, in the sense of measure theory, and that its generated σ-algebra $\sigma(Co(\Omega_B))$ is precisely the Baire σ-algebra $Ba(\Omega_B)$. A still larger σ-algebra is the family $Bo(\Omega_B)$ of Borel sets. For the remainder of this chapter we wish to discuss the problem of determining when the Stone map $Q : Co(\Omega_B) \longrightarrow \mathcal{B}$ can be extended, as a B.a. homomorphism (still with values in $\mathcal{B}$), to either $Ba(\Omega_B)$ or $Bo(\Omega_B)$.

Recall that a subset U of a topological space Ω is called *nowhere dense* if it has empty interior, i.e. $U^\circ = \emptyset$. A subset $V \subseteq \Omega$ is said to be of *first category* (or *meager*) if V is the union of at most countably many nowhere dense sets. An open subset U of Ω is called *regular* if it coincides with the interior of its closure , i.e. $U = (\overline{U})^\circ$. The collection of all regular open subsets of a topological space is an abstractly complete B.a. when partially ordered by set inclusion; see [16; Theorem 8.2]. Moreover, given a family $\{U_\alpha\}$ of regular open sets we have $\vee_\alpha U_\alpha = \overline{(\cup_\alpha U_\alpha)}^\circ$ and $\wedge_\alpha U_\alpha = (\cap_\alpha U_\alpha)^\circ$.

Exercise 26. Let Ω be a compact Hausdorff space.

(a) Show that every element of $Co(\Omega)$ is a regular open set.

(b) If, in addition, Ω is extremely disconnected, show that every regular open set belongs to $Co(\Omega)$.

(c) Let $\mathcal{B}$ be an abstractly complete B.a.. Show that $\mathcal{B}$ is isomorphic to the B.a. of all regular open subsets of some extremely disconnected, compact Hausdorff space. ∎

Let $\mathcal{A}$ and $\mathcal{B}$ be abstractly σ-complete B.a. 's and $\Phi : \mathcal{A} \longrightarrow \mathcal{B}$ be a B.a. homomorphism (c.f. Definition II.5). Then Φ is called a *σ-homomorphism* if $\Phi(\vee_n a_n) = \vee_n \Phi(a_n)$ for every countable set $\{a_n\}_{n\in\mathbb{N}} \subseteq \mathcal{A}$.

Definition II.10. Let $\mathcal{B}$ be a Boolean algebra.

(i) A non-empty collection $\mathcal{F}$ of elements from $\mathcal{B}$ is called an *ideal* if

(a) $u \vee v \in \mathcal{F}$ whenever $u, v \in \mathcal{F}$, and

(b) $u \in \mathcal{F}$ whenever $u \in \mathcal{B}$ and $u \leq v$ for some $v \in \mathcal{F}$.

(ii) Suppose that $\mathcal{B}$ is abstractly σ-complete . Then an ideal $\mathcal{F}$ in $\mathcal{B}$ is called a *σ-ideal* if $\vee_n u_n \in \mathcal{F}$ for all countable sets $\{u_n\}_{n\in\mathbb{N}} \subseteq \mathcal{F}$. ∎

Example 16. Let Ω be a compact Hausdorff space. Let $\mathcal{B} := 2^\Omega$. Using the fact that the union of two sets of first category is again of first category and that a subset of a set of first category is again of first category, it is clear that the collection of all subsets of first category is an ideal in $\mathcal{B}$. Since $\mathcal{B}$ is a σ-algebra it follows that $\vee_n A_n = \cup_{n=1}^\infty A_n$, for any countable collection $\{A_n\}_{n\in\mathbb{N}} \subseteq \mathcal{B}$. It is routine to check that the countable union of sets of first category is also of first category and hence, the collection of all subsets of first category is a σ-ideal in $\mathcal{B}$.

Let $\mathcal{M}_{Ba(\Omega)}$ denote the collection of all elements from $Ba(\Omega)$ which are of first category and $\mathcal{M}_{Bo(\Omega)}$ denote the collection of all elements from $Bo(\Omega)$ which are of first category. Since both $Ba(\Omega)$ and $Bo(\Omega)$ are sub-σ-algebras of $\mathcal{B}$ it is clear that $\mathcal{M}_{Ba(\Omega)}$ and $\mathcal{M}_{Bo(\Omega)}$

are σ-ideals in $Ba(\Omega)$ and $Bo(\Omega)$, respectively.   ∎

Let $\mathcal{F}$ be an ideal in a B.a. $\mathcal{B}$. We define an equivalence relation $\sim$ in $\mathcal{B}$ by $a \sim b$ if $a + b \in \mathcal{F}$; see (1) for the definition of $+$ in $\mathcal{B}$. The coset of $b \in \mathcal{B}$ is denoted by $[b]$. Then the set of all cosets $\mathcal{B}/\mathcal{F} := \{[b] : b \in \mathcal{B}\}$ becomes a B.a. with respect to the operations

$$[a] \vee [b] := [a \vee b], \quad [a] \wedge [b] := [a \wedge b] \quad \text{and} \quad [a]' := [a'], \qquad a, b \in \mathcal{B}.$$

The B.a. $\mathcal{B}/\mathcal{F}$ is called the *quotient* of $\mathcal{B}$ modulo $\mathcal{F}$. The map $h : \mathcal{B} \longrightarrow \mathcal{B}/\mathcal{F}$ defined by $h(b) = [b]$, for $b \in \mathcal{B}$, is a surjective B.a. homomorphism of $\mathcal{B}$ onto $\mathcal{B}/\mathcal{F}$. On the other hand, given B.a.'s $\mathcal{A}$ and $\mathcal{B}$ and a surjective B.a. homomorphism $h : \mathcal{A} \longrightarrow \mathcal{B}$, its *kernel* $\ker(h) := \{a \in \mathcal{A} : h(a) = 0_\mathcal{B}\}$ is an ideal in $\mathcal{A}$ and the map $g : \mathcal{A}/\ker(h) \longrightarrow \mathcal{B}$ given by $g([a]) = h(a)$ is a B.a. isomorphism onto $\mathcal{B}$. All of these notions and facts on quotient B.a.'s can be found in [40; §10], for example.

The following important extension result concerning the Stone map is known as the *Loomis-Sikorski theorem* ; see [16; Theorem 18.3], [26; Theorem 13] or [40; §29], for instance.

Theorem II.4.   *Let $\mathcal{B}$ be an abstractly σ-complete B.a. , $\Omega_\mathcal{B}$ be its Stone space and $\mathcal{M}_{Ba(\Omega_\mathcal{B})}$ be the σ-ideal of all Baire sets of first category . Then the Stone map $Q : Co(\Omega_\mathcal{B}) \longrightarrow \mathcal{B}$ has a unique extension to a σ-homomorphism $\overline{Q} : Ba(\Omega_\mathcal{B}) \longrightarrow \mathcal{B}$ with kernel $\ker(\overline{Q}) = \mathcal{M}_{Ba(\Omega_\mathcal{B})}$.*

Remark.   (a) The *Baire category theorem* for compact spaces states that every set of first category in a compact Hausdorff space has empty interior , [16; Theorem 18.2]. It follows that if $E, F \in Co(\Omega_\mathcal{B})$ satisfy $E \sim F$ modulo $\mathcal{M}_{Ba(\Omega_\mathcal{B})}$, then actually $E = F$. Accordingly, for each $E \in Ba(\Omega_\mathcal{B})$ there is a *unique* set $\widetilde{E} \in Co(\Omega_\mathcal{B})$ such that the symmetric difference $E \triangle \widetilde{E}$ (which corresponds to $E + \widetilde{E}$ in $Ba(\Omega_\mathcal{B})$) belongs to $\mathcal{M}_{Ba(\Omega_\mathcal{B})}$. The extension $\overline{Q}$ of Q is then given by $\overline{Q}(E) := Q(\widetilde{E})$.

(b) The uniqueness of the extension $\overline{Q}$ can be argued as follows. Let $P : Ba(\Omega_\mathcal{B}) \longrightarrow \mathcal{B}$ be another σ-homomorphism which coincides with Q on $Co(\Omega_\mathcal{B})$. Define $\mathcal{A} := \{E \in Ba(\Omega_\mathcal{B}) : P(E) = \overline{Q}(E)\}$. By hypothesis $Co(\Omega_\mathcal{B}) \subseteq \mathcal{A}$. Let $\{E_n\}_{n=1}^\infty \subseteq \mathcal{A}$ be increasing. Since $Ba(\Omega_\mathcal{B})$ is a σ-algebra we have $\vee_n E_n = \cup_{n=1}^\infty E_n$. Since $P, \overline{Q}$ are σ-homomorphisms it follows that

$$\overline{Q}(\cup_{n=1}^\infty E_n) = \vee_n \overline{Q}(E_n) = \vee_n P(E_n) = P(\cup_{n=1}^\infty E_n).$$

Accordingly, $\cup_{n=1}^\infty E_n \in \mathcal{A}$. A similar argument shows that $\cap_{n=1}^\infty F_n \in \mathcal{A}$ for every decreasing family $\{F_n\}_{n=1}^\infty \subseteq \mathcal{A}$. So, $\mathcal{A}$ is a *monotone class* of sets (c.f. [38; Chapter 7] for the definition) containing the *algebra of sets* $Co(\Omega_\mathcal{B})$, from which it follows that $\mathcal{A} = \sigma(Co(\Omega_\mathcal{B}))$. Then Proposition II.5 shows that $\mathcal{A} = Ba(\Omega_\mathcal{B})$, that is, $P = \overline{Q}$.   ∎

Under extra restrictions on $\mathcal{B}$ the map $\overline{Q}$ has a further extension.

Theorem II.5.   *Let $\mathcal{B}$ be an abstractly complete B.a., $\Omega_\mathcal{B}$ be its Stone space and $\mathcal{M}_{Bo(\Omega_\mathcal{B})}$ be the σ-ideal of all Borel sets of first category . Then the Stone map $Q : Co(\Omega_\mathcal{B}) \longrightarrow \mathcal{B}$ has an extension to a σ-homomorphism $\widehat{Q} : Bo(\Omega_\mathcal{B}) \longrightarrow \mathcal{B}$ with the following properties.*

(a) $\ker(\widehat{Q}) = \mathcal{M}_{Bo(\Omega_\mathcal{B})}$.

(b) $\vee_\alpha \widehat{Q}(V_\alpha) = \widehat{Q}(\cup_\alpha V_\alpha)$ *for every family* $\{V_\alpha\}$ *of open sets in* $\Omega_\mathcal{B}$.

(c) *For each open set $E \subseteq \Omega_B$ we have*

$$\widehat{Q}(E) = \vee\{\widehat{Q}(F) : \ F \text{ closed, } F \subseteq E\}.$$

(d) *If $P : Bo(\Omega_B) \longrightarrow \mathcal{B}$ is another σ-homomorphism which coincides with Q on $Co(\Omega_B)$, then P and $\widehat{Q}$ agree on the open subsets of Ω_B and on the Baire σ-algebra $Ba(\Omega_B)$.*

Proof. Since Ω_B is extremely disconnected (c.f. Proposition II.4) it follows from Exercise 26 that the regular open subsets of Ω_B are precisely the elements of $Co(\Omega_B)$. Hence the existence of $\widehat{Q}$, that it is a σ-homomorphism, and property (a) are well known; see [26; Theorem 4] or [16; Theorem 21.7], for example.

(b) Let $V = \cup_\alpha V_\alpha$. Since Q is surjective , there are sets $U_1, U_2 \in Co(\Omega_B)$ for which $Q(U_1) = \vee_\alpha \widehat{Q}(V_\alpha)$ and $Q(U_2) = \widehat{Q}(V)$. Since $V_\alpha \subseteq V$ for each α and $\widehat{Q}$ is a B.a. homomorphism, we have $Q(U_1) \leq Q(U_2)$. Conversely, let $\{W_\beta\}$ be a subfamily of $Co(\Omega_B)$ such that W_β is contained in some V_α and $V = \cup_\beta W_\beta$; this is possible as $Co(\Omega_B)$ forms a base for the topology in Ω_B. Given any β we have, for some α, that $Q(W_\beta) \leq \widehat{Q}(V_\alpha) \leq Q(U_1)$ and so $W_\beta \subseteq U_1$ as Q is a B.a. isomorphism. Consequently, $V = \cup_\beta W_\beta \subseteq U_1$ and $Q(U_2) = \widehat{Q}(V) \leq Q(U_1)$. Thus $Q(U_1) = Q(U_2)$ which establishes (b).

(c) Since $Co(\Omega_B)$ forms a base for the topology in Ω_B there is a family of sets $\{H_\beta\} \subseteq Co(\Omega_B)$ whose union is E. By part (b)

$$\widehat{Q}(E) = \vee_\beta Q(H_\beta) \leq \vee\{\widehat{Q}(F) : \ F \subseteq E, F \text{ closed}\} \leq \widehat{Q}(E).$$

This establishes (c).

(d) The formula in (b) also hold with P in place of $\widehat{Q}$. So, if $V \subseteq \Omega_B$ is an open set, then the fact that $Co(\Omega_B)$ forms a base for the topology in Ω_B implies that

$$P(V) = \vee\{Q(W) : V \supseteq W \in Co(\Omega_B)\} = \widehat{Q}(V).$$

That P and $\widehat{Q}$ also agree on $Ba(\Omega_B)$ can be argued along the lines indicated in Remark (b) after Theorem II.4. ∎

Property (c) is a kind of regularity condition akin to that seen in measure theory. The question arises of whether (c) holds for arbitrary Borel sets E rather than just open sets? Unfortunately, this is not the case in general.

Example 17. Let $\mathcal{M}_{Bo(\mathbb{R})}$ denote the σ-ideal in $Bo(\mathbb{R})$ consisting of all sets of first category . Then the quotient B.a. $\mathcal{B} := Bo(\mathbb{R})/\mathcal{M}_{Bo(\mathbb{R})}$ is abstractly complete , uncountable, *atomless* and has a countable dense set (in the sense of B.a. 's), [40; p.94]. So, its Stone space Ω_B is *separable* in the topological sense, meaning it has a countable dense set, say $D = \{d_n : n \in \mathbb{N}\}$. Since each singleton set $\{d_n\}$ is closed in Ω_B, it must have empty interior . Otherwise, $\{d_n\} \in Co(\Omega_B)$ in which case $\{d_n\}$ would be an atom of $Co(\Omega_B)$ and hence, $Q(\{d_n\})$ would be an atom of $\mathcal{B}$. So, $D \in \mathcal{M}_{Bo(\Omega_B)}$ and hence $\widehat{Q}(D) = 0_\mathcal{B}$, by Theorem II.5(a). If (c) of Theorem II.5 was true for arbitrary sets $E \in Bo(\Omega_B)$ we would have

(4) $$\widehat{Q}(D^c) = \vee\{\widehat{Q}(F) : \ F \text{ closed, } F \subseteq D^c\}.$$

But, if F is any closed set satisfying $F \subseteq D^c$, then F^c is an open set containing the dense set D and so $\widehat{Q}(F^c) = 1_B$. Accordingly, $\widehat{Q}(F) = 0_B$ and so the right-hand-side of (4) equals 0_B. However, the left-hand-side of (4) equals 1_B. So, (4) cannot be an equality. ∎

Chapter III

Boolean algebras of projections and uniformly closed operator algebras

Let X be a Banach space and $\mathcal{B} \subseteq \mathcal{L}(X)$ be a B.a. of projections which is bounded. The aim of this chapter is to identify the closed subalgebra $\langle \mathcal{B} \rangle_u^-$ generated by $\mathcal{B}$ with respect to the *operator norm topology* in $\mathcal{L}(X)$. Since $\langle \mathcal{B} \rangle_u^-$ is a commutative, unital Banach algebra it is possible to apply the general methods of Banach algebras. Indeed, this is the approach adopted in [15; Chapter XVII, §2]. However, our approach will be via the Stone space $\Omega_\mathcal{B}$ of $\mathcal{B}$ and the methods of B.a.'s as developed in the previous chapter; see also [13] and [33] for this approach. The reason for this approach is that it gives a consistency of treatment throughout the text, since B.a. methods and spectral measures *must* be used in the next chapters when considering the closed algebra generated by $\mathcal{B}$ with respect to other *non-normable* topologies on $\mathcal{L}(X)$, where Banach algebras no longer play an effective role. The basic idea in this chapter is to realize $\mathcal{B}$ as the range of a *finitely additive* spectral measure ; the boundedness of $\mathcal{B}$ enters to ensure that at least bounded measurable functions are "integrable" with respect to this finitely additive spectral measure (via a suitable extension process from the simple functions). It turns out that $\langle \mathcal{B} \rangle_u^-$ is isomorphic to $C(\Omega_\mathcal{B})$. Moreover, there exists a bounded B.a. of projections $\mathcal{A}$ in $\mathcal{L}(X')$, indexed by the elements of $Bo(\Omega_\mathcal{B})$, such that $\mathcal{B}' := \{F' : F \in \mathcal{B}\} \subseteq \mathcal{A}$ and the isomorphism $\Phi : C(\Omega_\mathcal{B}) \longrightarrow \langle \mathcal{B} \rangle_u^-$ has the property that each dual operator $\Phi(f)'$, for $f \in C(\Omega_\mathcal{B})$, is given as the integral of f against an $\mathcal{L}(X')$-valued ("almost") spectral measure taking its values in $\mathcal{A}$. Without any further restrictions on X or $\mathcal{B}$ this is the best that can be expected.

Let X be a Banach space. Then $\mathcal{L}(X)$ is also a Banach space for the *operator norm*

$$\|T\| := \sup\{\|Tx\| : x \in X, \|x\| \le 1\}, \qquad T \in \mathcal{L}(X).$$

This norm topology on $\mathcal{L}(X)$ is called the *uniform operator topology* ; if we wish to stress this topology is being considered we will write $\mathcal{L}_u(X)$. A subspace of $\mathcal{L}(X)$ is called closed for the uniform operator topology if it is closed in $\mathcal{L}_u(X)$. The inequality

$$\|TS\| \le \|T\| \cdot \|S\|, \qquad T, S \in \mathcal{L}(X),$$

shows that $\mathcal{L}(X)$ is a *Banach algebra* : it has the identity operator I as its unit. If X is at least 2-dimensional, then $\mathcal{L}_u(X)$ is not commutative, that is, there exist operators $S, T \in \mathcal{L}(X)$ such that $ST \neq TS$.

An operator $T \in \mathcal{L}(X)$ is called *invertible* if there exists $S \in \mathcal{L}(X)$, necessarily unique, such that $ST = I = TS$. We denote S by T^{-1}. Let $\mathrm{Inv}(X)$ denote the set of all invertible operators in $\mathcal{L}(X)$.

Lemma III.1. *Let X be a Banach space.*

(a) *If $T \in \mathcal{L}(X)$ satisfies $\|T\| < 1$, then $(I - T) \in \mathrm{Inv}(X)$.*

(b) *$\mathrm{Inv}(X)$ is an open subset of $\mathcal{L}_u(X)$.*

Proof.   (a) Since the geometric series $\sum_{n=0}^{\infty} \|T\|^n < \infty$ and $\sum_{n=0}^{\infty} \|T^n\| \leq \sum_{n=0}^{\infty} \|T\|^n$, it follows that the series $\sum_{n=0}^{\infty} T^n$ is absolutely convergent in $\mathcal{L}_u(X)$ to some operator $S \in \mathcal{L}(X)$, say. Since

$$(I - T) \cdot \sum_{j=0}^{n} T^j = I - T^{n+1} = \left(\sum_{j=0}^{n} T^j \right) \cdot (I - T), \qquad n \in \mathbb{N},$$

by letting $n \to \infty$ and noting that $\|T^{n+1}\| \leq \|T\|^{n+1} \longrightarrow 0$ we see that $(I - T)S = I = S(I - T)$. Hence, $(I - T) \in \mathrm{Inv}(X)$ with $(I - T)^{-1} = S = \sum_{n=0}^{\infty} T^n$.

(b) Fix $S \in \mathrm{Inv}(X)$. Let $R \in \mathcal{L}(X)$ satisfy $\|R - S\| < \frac{1}{\|S^{-1}\|}$. Then

$$\|I - RS^{-1}\| = \|(S - R)S^{-1}\| \leq \|S - R\| \cdot \|S^{-1}\| < 1$$

and so, by part (a) applied to $T = I - RS^{-1}$, we see that $RS^{-1} = (I - T) \in \mathrm{Inv}(X)$. Hence, $R = (RS^{-1})S$ is also invertible being the product of invertible operators. This shows that the open ball in $\mathcal{L}_u(X)$ with centre S and radius $\frac{1}{\|S^{-1}\|}$ is contained in $\mathrm{Inv}(X)$. Since $S \in \mathrm{Inv}(X)$ is arbitrary we are done.   ■

Exercise 27.   Let X be a Banach space. Show that the map $T \mapsto T^{-1}$ is a continuous map of $\mathrm{Inv}(X)$ onto $\mathrm{Inv}(X)$ with respect to the *operator norm topology* inherited from $\mathcal{L}_u(X)$. ■

The following inequalities will be needed later.

Exercise 28.[∗]   Let X be a Banach space with norm $\| \cdot \|$.

(a) Show that $\big| \|x\| - \|y\| \big| \leq \|x - y\|$, for all $x, y \in X$.

(b) Deduce that $\|x - y\| \geq \|x\| - \|y\|$, for all $x, y \in X$.   ■

We now turn our attention to a special class of B.a. 's.

Definition III.1. Let X be a Banach space.

(a) A collection $\mathcal{B} \subseteq \mathcal{L}(X)$ of *commuting* projection operators is called a *B.a. of projections* if $0, I \in \mathcal{B}$ and if it is a B.a. with respect to the partial order $\leq$ defined by $E \leq F$ if $EF = E = FE$.

(b) A B.a. of projections $\mathcal{B} \subseteq \mathcal{L}(X)$ is called *bounded* if

$$\|\mathcal{B}\| := \sup \{ \|E\| : E \in \mathcal{B} \} < \infty.$$ ■

The notation E' for the B.a. complement in $\mathcal{B}$ of an element $E \in \mathcal{B}$ (so, in particular, $E' \in \mathcal{L}(X)$) should not be confused with the dual operator $E' \in \mathcal{L}(X')$. It will always be clear from the context which of the two distinct notions is meant by E'.

Exercise 29.   Let X be a Banach space and $B \subseteq L(X)$ be a B.a. of projections.

(a) For elements $E, F \in B$ show that $E \leq F$ if and only if $EX \subseteq FX$.

(b) Show that the B.a. operations $\wedge, \vee$ and complementation in B are given by $E \wedge F = EF$ and $E \vee F = E + F - EF$ and $E' = I - E$ (*Hint:* See Exercise 14). ∎

Exercise 30.[*]   (a) Let $X = \mathbb{C}^2$ and consider the matrices $A = \left(\begin{smallmatrix} 1 & 0 \\ 0 & 0 \end{smallmatrix} \right)$ and $B = \left(\begin{smallmatrix} 0 & 0 \\ 0 & 1 \end{smallmatrix} \right)$, interpreted as elements of $L(X)$. Show that both A and B are projections and satisfy $AB = BA$. Show that $(B - A)$ is *not* a projection.

(b) Let X be a Banach space and $B \subseteq L(X)$ be a B.a. of projections. Show that if $A, B \in B$ satisfy $A \leq B$, then $(B - A) \in B$. ∎

We will be dealing exclusively with *bounded* B.a.'s of projections. The following example shows that not all B.a.'s of projections are bounded.

Example 18.   Let $X = L^p(\mathbb{R})$, for some $p \in (1, 2)$; see [38; Chapter 3]. For each $t \in \mathbb{R}$, define the translation operator $T_t \in L(X)$ by $T_t f = f_t$, for $f \in X$, where $f_t(s) = f(s + t)$ for a.e. $s \in \mathbb{R}$. A projection $E \in L(X)$ is called a *p-multiplier projection* if $ET_t = T_t E$, for all $t \in \mathbb{R}$. It is a known fact from harmonic analysis that the family B_p of all p-multiplier projections is a B.a. of projections for which $\sup\{\|E\| : E \in B_p\} = \infty$. If $p = 1$, then $B_1 = \{0, I\}$ and if $p = 2$, then B_2 consists of selfadjoint projections. So, both B_1 and B_2 are bounded B.a.'s of projections. ∎

The following remarkable result, due to W.G. Bade [1], shows that a large class of B.a.'s of projections are always bounded. The proof given below is an expanded version of Bade's original proof.

Theorem III.1.   *Let X be a Banach space. Then every abstractly σ-complete B.a. of projections in $L(X)$ is necessarily bounded.*

Proof.   Proceeding by contradiction, assume that there exists an abstractly σ-complete B.a. of projections $B \subseteq L(X)$ which is *not* bounded. Declare a projection $E \in B$ to have *property* (α) if

$$\alpha(E) := \sup\{\|F\| : F \in B, F \leq E\} = \infty.$$

Suppose there exists $E \in B$ such that both E and $E' := (I - E) \in L(X)$ do *not* have property (α). Since every $F \in B$ satisfies $F = FE + FE'$ with $FE \leq E$ and $FE' \leq E'$ it follows that $\|F\| \leq \|FE\| + \|FE'\| \leq \alpha(E) + \alpha(E')$. Since $F \in B$ is arbitrary we contradict our assumption that B is not bounded. So, for every $E \in B$, at least one of E or E' must have property (α).

Suppose that $E \in B$ has property (α) and $F \in B$ satisfies $F \leq E$, in which case $(E - F) \in B$ by Exercise 30(b). Suppose that both F and $(E - F)$ do *not* have property (α). Since every $H \in B$ with $H \leq E$ satisfies $H = HE = HF + H(E - F)$, where $HF \leq F$ and $H(E - F) \leq (E - F)$ it follows that $\|H\| \leq \|HF\| + \|H(E - F)\| \leq \alpha(F) + \alpha(E - F)$. Accordingly, $\alpha(E) < \infty$ contrary to the choice of E. Hence, if $E \in B$ has property (α) and $F \leq E$, with $F \in B$, then at least one of F or $(E - F)$ also has property (α).

From the above discussion there exists some element $E_1 \in B$ with property (α). Then, by definition of $\alpha(E_1) = \infty$, there is $F_1 \in \mathcal{B}$ with $F_1 \leq E_1$ such that $\|F_1\| \geq 2 + 2\|E_1\|$, i.e.

$$(1) \qquad\qquad \|F_1\| - \|E_1\| \geq 2 + \|E_1\|.$$

Let E_2 be any element from $\{F_1, (E_1 - F_1)\}$ having property (α). If $E_2 = F_1$, then certainly $E_2 \leq E_1$. On the other hand if $E_2 = (E_1 - F_1)$, then

$$E_2 E_1 = (E_1 - F_1)E_1 = E_1 - F_1 E_1 = E_1 - F_1 = E_2$$

and hence again $E_2 \leq E_1$. So, $E_2 \in \mathcal{B}$ satisfies $E_2 \leq E_1$. Moreover, if $E_2 = F_1$, then

$$\|E_2\| = \|F_1\| \geq 2 + 2\|E_1\| \geq 2 + \|E_1\|.$$

On the other hand if $E_2 = (E_1 - F_1)$, then (1) and Exercise 28(b) imply that

$$\|E_2\| = \|E_1 - F_1\| \geq \|F_1\| - \|E_1\| \geq 2 + \|E_1\|.$$

So, we have produced an element $E_2 \in \mathcal{B}$ satisfying both $E_2 \leq E_1$ and $\|E_2\| \geq 2 + \|E_1\|$.

Since E_2 has property (α) there is $F_2 \in \mathcal{B}$ with $F_2 \leq E_2$ such that $\|F_2\| \geq 3 + 2\|E_2\|$. Let E_3 be any element from $\{F_2, (E_2 - F_2)\}$ with property (α). Then again $E_3 \in \mathcal{B}$ satisfies $E_3 \leq E_2$ and a similar argument as above shows that $\|E_3\| \geq 3 + \|E_2\|$. Proceeding inductively produces a monotone decreasing sequence $\{E_n\}_{n=1}^{\infty} \subseteq \mathcal{B}$ satisfying

$$(2) \qquad\qquad \|E_n\| \geq n + \|E_{n-1}\|, \qquad n \geq 2.$$

Define $G_n := (E_n - E_{n+1})$ for $n \in \mathbb{N}$. Since $E_{n+1} \leq E_n$ it is clear that each $G_n \in \mathcal{B}$; see Exercise 30(b). Suppose that $n > m$, in which case $E_n E_m = E_n$ and $E_{n+1} E_m = E_{n+1}$ and $E_{n+1} E_{m+1} = E_{n+1}$ and $E_n E_{m+1} = E_n$. Using these identities it follows, by expanding the right-hand-side of the expression $G_n G_m = (E_n - E_{n+1})(E_m - E_{m+1})$, that $G_n G_m = 0$. Hence, the projections $\{G_n\}_{n=1}^{\infty}$ are *pairwise disjoint* in $\mathcal{B}$. Moreover, by (2) and Exercise 28(b), $\|G_n\| = \|E_n - E_{n+1}\| \geq \|E_{n+1}\| - \|E_n\| \geq (n+1)$, that is, $\lim_{n \to \infty} \|G_n\| = \infty$.

By selecting subsequences from $\{G_n\}_{n=1}^{\infty}$, a collection of mutually disjoint sequences of pairwise disjoint projections $\{H_{j,k}\}_{k=1}^{\infty}$, for each $j \in \mathbb{N}$, is obtained (all from $\mathcal{B}$) such that

$$(3) \qquad\qquad \lim_{k \to \infty} \|H_{j,k}\| = \infty, \qquad j \in \mathbb{N}.$$

Since $\mathcal{B}$ is abstractly σ-complete the elements $P_j := \vee_{k=1}^{\infty} H_{j,k}$ exist in $\mathcal{B}$, for each $j \in \mathbb{N}$. For $m \neq n$ it follows from Exercise 16(b) that

$$(4) \qquad P_n P_m = P_n \wedge P_m = P_n \wedge (\vee_{k=1}^{\infty} H_{m,k}) = \vee_{k=1}^{\infty}(P_n \wedge H_{m,k}).$$

But, since $H_{m,k} H_{n,r} = 0$, for all $k, r \in \mathbb{N}$ and $m \neq n$, it follows that

$$H_{m,k} \wedge P_n = H_{m,k} \wedge (\vee_{r=1}^{\infty} H_{n,r}) = \vee_{r=1}^{\infty}(H_{m,k} \wedge H_{n,r}) = \vee_{r=1}^{\infty}(H_{m,k} H_{n,r}) = 0,$$

for all $k \in \mathbb{N}$. It follows from (4) that $P_n P_m = 0$ whenever $m \neq n$.

Fix m and $x \neq 0$. Let $X_m := P_m X$. Suppose that $P_m x \neq 0$. Since $H_{m,n} = H_{m,n} P_m$, for $n \in \mathbb{N}$, we have

$$\frac{\|H_{m,n} x\|}{\|x\|} = \frac{\|H_{m,n} P_m x\|}{\|x\|} = \frac{\|P_m x\|}{\|x\|} \frac{\|H_{m,n} P_m x\|}{\|P_m x\|}.$$

But, $\frac{\|P_m x\|}{\|x\|} \leq \|P_m\|$ and $\frac{\|H_{m,n} P_m x\|}{\|P_m x\|} \leq \|H_{m,n}\|_{\mathcal{L}(X_m)}$, and so $\frac{\|H_{m,n} x\|}{\|x\|} \leq \|P_m\| \cdot \|H_{m,n}\|_{\mathcal{L}(X_m)}$. In the case when $P_m x = 0$ we have

$$\frac{\|H_{m,n} x\|}{\|x\|} = \frac{\|H_{m,n} P_m x\|}{\|x\|} = 0 \leq \|P_m\| \cdot \|H_{m,n}\|_{\mathcal{L}(X_m)}.$$

So, we conclude that

$$\|H_{m,n}\| \leq \|P_m\| \cdot \|H_{m,n}\|_{\mathcal{L}(X_m)}, \qquad m, n \in \mathbb{N}.$$

It follows from this inequality and (3) that, for each *fixed* $m \in \mathbb{N}$, we have

$$\lim_{n \to \infty} \|H_{m,n}\|_{\mathcal{L}(X_m)} = \infty.$$

So, we can select an increasing sequence $\{n_k\}_{k=1}^{\infty} \subseteq \mathbb{N}$ and unit vectors $x_k \in X_k$ such that $\|H_{k,n_k} x_k\| > k$, for all $k \in \mathbb{N}$. Let $Q = \vee_{k=1}^{\infty} H_{k,n_k}$. Since $P_k x_k = x_k$ (as $x_k \in X_k$) and $Q P_\ell = H_{\ell, n_\ell}$, for all $k, \ell \in \mathbb{N}$ we have $\|Q x_k\| = \|Q P_k x_k\| = \|H_{k,n_k} x_k\| > k$, for $k \in \mathbb{N}$. It follows that $\infty = \sup\{\|Q x_k\| : k \in \mathbb{N}\} \leq \|Q\|$ which is impossible since $Q \in \mathcal{L}(X)$ means that $\|Q\| < \infty$. Hence, $\mathcal{B}$ must be bounded. ∎

Let $\{A_\alpha : \alpha \in A\}$ be a family of subalgebras of $\mathcal{L}(X)$, that is, each A_α is a vector subspace of $\mathcal{L}(X)$ and $TS \in A_\alpha$ whenever $T, S \in A_\alpha$. Then the intersection $\cap_\alpha A_\alpha$ is also a subalgebra of $\mathcal{L}(X)$. So, for any subset $\mathcal{M} \subseteq \mathcal{L}(X)$ there is a *smallest* subalgebra in $\mathcal{L}(X)$, denoted by $\langle \mathcal{M} \rangle$, which contains $\mathcal{M}$, namely the intersection of all subalgebras containing $\mathcal{M}$. We call $\langle \mathcal{M} \rangle$ the subalgebra of $\mathcal{L}(X)$ *generated by* $\mathcal{M}$. The *closed* algebra generated by $\mathcal{M}$ is the smallest closed subalgebra of $\mathcal{L}_u(X)$ containing $\mathcal{M}$; it is, of course, the closure of $\langle \mathcal{M} \rangle$ in $\mathcal{L}_u(X)$ and is denoted by $\langle \mathcal{M} \rangle_u^-$. We call $\langle \mathcal{M} \rangle_u^-$ the *uniformly closed algebra generated by* $\mathcal{M}$.

The following result gives a complete description of $\langle \mathcal{B} \rangle_u^-$ in the case when $\mathcal{B}$ is a bounded B.a. of projections. A proof based on the theory of Banach algebras can be found in [15; Chapter XVII, §2]. We have decided to give a proof based on the theory of Stone spaces for B.a.'s as this is the underlying approach of the entire text; see also [13; Proposition 5.43]. The crucial point turns out to be the fact that every bounded finitely additive spectral measure $P : \Sigma \longrightarrow \mathcal{L}(X)$ defined on an algebra of sets Σ (see Definitions III.2 and III.6 below) yields a continuous homomorphism (via integration) of $B^{\infty}(\Sigma)$ into $\mathcal{L}_u(X)$.

Theorem III.2. *Let X be a Banach space and $\mathcal{B} \subseteq \mathcal{L}(X)$ be a bounded B.a. of projections.*

(a) *$\langle \mathcal{B} \rangle_u^-$ is inverse closed in $\mathcal{L}_u(X)$, that is, $T^{-1} \in \langle \mathcal{B} \rangle_u^-$ whenever $T \in \langle \mathcal{B} \rangle_u^- \cap \text{Inv}(X)$.*

(b) *$\langle \mathcal{B} \rangle_u^-$ is isomorphic (as a commutative, unital Banach algebra) to $C(\Omega_{\mathcal{B}})$, where $\Omega_{\mathcal{B}}$ is the Stone space of $\mathcal{B}$, via an isomorphism $\Phi : C(\Omega_{\mathcal{B}}) \longrightarrow \langle \mathcal{B} \rangle_u^-$ which satisfies*

$$\Phi(\chi_E) = Q(E), \qquad E \in Co(\Omega_{\mathcal{B}}),$$

where $Q : Co(\Omega_{\mathcal{B}}) \longrightarrow \mathcal{B}$ is the Stone map , and the inequalities

$$\|f\|_\infty \leq \|\Phi(f)\| \leq 4\|\mathcal{B}\| \cdot \|f\|_\infty, \qquad f \in C(\Omega_{\mathcal{B}}).$$

Proof.   (a) Note that $\langle \mathcal{B} \rangle$ consists of all operators of the form

$$(5) \qquad\qquad\qquad\qquad U = \sum_{j \in \mathcal{F}} \alpha_j F_j$$

where $\mathcal{F} \subseteq \mathbb{N}$ is any finite set, $\{\alpha_j : j \in \mathcal{F}\}$ is any set of distinct complex numbers and

$(6) \quad \{F_j : j \in \mathcal{F}\} \subseteq \mathcal{B}$ are non-zero projections with $F_j F_k = 0$, $j \neq k$, and $\displaystyle\sum_{j \in \mathcal{F}} F_j = I$.

Let $T \in \langle \mathcal{B} \rangle_u^- \cap \mathrm{Inv}(X)$. Choose $\{U_n\}_{n=1}^\infty \subseteq \langle \mathcal{B} \rangle$ such that $U_n \to T$ in $\mathcal{L}_u(X)$. Then by Lemma III.1(b) there is N such that $U_n \in \mathrm{Inv}(X)$ for all $n \geq N$. Exercise 27 shows that $U_n^{-1} \longrightarrow T^{-1}$ as $n \to \infty$. So, T^{-1} will belong to $\langle \mathcal{B} \rangle_u^-$ provided we know that $U_n^{-1} \in \langle \mathcal{B} \rangle$ for all $n \geq N$.

So, suppose that $U \in \langle \mathcal{B} \rangle$ is given by (5) and (6) with $U \in \mathrm{Inv}(X)$. We claim that $\alpha_j \neq 0$ for all $j \in \mathcal{F}$. For, suppose that $\alpha_{j_0} = 0$ for some $j_0 \in \mathcal{F}$. Since $F_{j_0} \neq 0$, there is $x \neq 0$ with $F_{j_0} x = x$. Then (6) implies that

$$Ux = U F_{j_0} x = \Big(\sum_{j \in \mathcal{F}} \alpha_j F_j\Big) F_{j_0} x = \alpha_{j_0} F_{j_0} x = 0.$$

Hence, $x \in \ker(U)$ contradicting $U \in \mathrm{Inv}(X)$. So, $\alpha_j \neq 0$ for all $j \in \mathcal{F}$ and hence $U^{-1} = \sum_{j \in \mathcal{F}} (\frac{1}{\alpha_j}) F_j$, showing that $U^{-1} \in \langle \mathcal{B} \rangle$.

(b) Let $Q : Co(\Omega_{\mathcal{B}}) \longrightarrow \mathcal{B}$ be the Stone map in which case $Q(\emptyset) = 0$, $Q(\Omega) = I$ and $Q(A \cap B) = Q(A) Q(B)$. Moreover, by Remark (c) after Theorem II.2 and the B.a. operations in $\mathcal{B}$ we see that Q is finitely additive (in the sense of measures) on the algebra of sets $Co(\Omega_{\mathcal{B}})$, i.e. $Q(\cup_{j=1}^n E_j) = \sum_{j=1}^n Q(E_j)$ for all pairwise disjoint sets $\{E_j\}_{j=1}^n \subseteq Co(\Omega_{\mathcal{B}})$. So, Q has a *unique* extension to $\mathrm{sim}(Co(\Omega_{\mathcal{B}}))$, denoted by Φ, defined by linearity and the property $\Phi(\chi_E) := Q(E)$ for $E \in Co(\Omega_{\mathcal{B}})$.

Suppose that $f = \sum_{j=1}^n \alpha_j \chi_{E_j} \in \mathrm{sim}(Co(\Omega_{\mathcal{B}}))$ has its standard representation , in which case $\|f\|_\infty = \max_{1 \leq j \leq n} |\alpha_j|$. Then $\Phi(f) = \sum_{j=1}^n \alpha_j Q(E_j)$. Fix $x \in X$ with $\|x\| \leq 1$ and let $m := Qx$ be the X-valued, bounded finitely additive measure $E \mapsto Q(E)x$, for $E \in Co(\Omega_{\mathcal{B}})$. By (9) of Chapter I and Proposition I.2 we have

$$(7) \qquad \|\Phi(f)x\| \leq \|f\|_\infty \cdot \|m\|(\Omega_{\mathcal{B}}) \leq 4\|f\|_\infty \sup\{\|Q(E)x\| : E \in Co(\Omega_{\mathcal{B}})\},$$

from which it follows (using $\|Q(E)x\| \leq \|\mathcal{B}\| \cdot \|x\|$ and taking the supremum over all $x \in X$ satisfying $\|x\| \leq 1$) that $\|\Phi(f)\| \leq 4\|\mathcal{B}\| \cdot \|f\|_\infty$.

Now fix $j_0 \in \mathcal{F}$. Since $Q(E_{j_0}) \neq 0$ we can choose $x_{j_0} \in Q(E_{j_0})X$ with $\|x_{j_0}\| = 1$. Then, using the fact that $Q(E_j) Q(E_{j_0}) = 0$ whenever $j \neq j_0$, we have

$$\Phi(f)x = \sum_{j=1}^n \alpha_j Q(E_j) Q(E_{j_0}) x_{j_0} = \alpha_{j_0} Q(E_{j_0}) x_{j_0} = \alpha_{j_0} x_{j_0}$$

and so $\|\Phi(f)x_{j_0}\| = |\alpha_{j_0}|$. Accordingly, $\|\Phi(f)\| \geq \|\Phi(f)x_{j_0}\| = |\alpha_{j_0}|$. Since j_0 is arbitrary we deduce that $\|f\|_\infty = \max_{1 \leq j \leq n} |\alpha_j| \leq \|\Phi(f)\|$. So, we have established that

$$(8) \qquad \|f\|_\infty \leq \|\Phi(f)\| \leq 4\|\mathcal{B}\| \cdot \|f\|_\infty, \quad f \in \mathrm{sim}(Co(\Omega_\mathcal{B})).$$

To see that $\Phi : \mathrm{sim}(Co(\Omega_\mathcal{B})) \longrightarrow \langle\mathcal{B}\rangle$ is also *multiplicative* note that if $f = \sum_{j=1}^n \alpha_j \chi_{E_j}$ has its standard representation and $E \in Co(\Omega_\mathcal{B})$, then

$$f\chi_E = \sum_{j=1}^n \alpha_j \chi_{E \cap E_j}$$

is the standard representation of $f\chi_E$ and so (using the fact Q is a B.a. isomorphism)

$$
\begin{aligned}
\Phi(f\chi_E) &= \sum_{j=1}^n \alpha_j \Phi(\chi_{E \cap E_j}) = \sum_{j=1}^n \alpha_j Q(E \cap E_j) = \\
&= \sum_{j=1}^n \alpha_j Q(E)Q(E_j) = [\sum_{j=1}^n \alpha_j Q(E_j)]Q(E) = \\
&= \Phi(f)\Phi(\chi_E).
\end{aligned}
$$

This formula and the linearity of Φ then imply that

$$(9) \qquad \Phi(fg) = \Phi(f)\Phi(g), \quad f, g \in \mathrm{sim}(Co(\Omega_\mathcal{B})).$$

It is clear from (8) and (9) that Φ has a unique continuous extension, which is still linear and multiplicative, from the closure of $\mathrm{sim}(Co(\Omega_\mathcal{B}))$, taken in $C(\Omega_\mathcal{B})$, into $\langle\mathcal{B}\rangle_u^-$. Still denoting this extension by Φ it is clear that

$$(10) \qquad \|f\|_\infty \leq \|\Phi(f)\| \leq 4\|f\|_\infty \|\mathcal{B}\|, \quad f \in C(\Omega_\mathcal{B}),$$

after using Proposition II.1 to establish that $\overline{\mathrm{sim}(Co(\Omega_\mathcal{B}))} = C(\Omega_\mathcal{B})$.

Since $\langle\mathcal{B}\rangle = \Phi(\mathrm{sim}(Co(\Omega_\mathcal{B}))$ it is clear from (10) that $\Phi(C(\Omega_\mathcal{B})) = \langle\mathcal{B}\rangle_u^-$, i.e. Φ maps $C(\Omega_\mathcal{B})$ *onto* $\langle\mathcal{B}\rangle_u^-$. Also (10) shows that Φ is *injective* on $C(\Omega_\mathcal{B})$. Hence, Φ is a Banach algebra isomorphism . ∎

Exercise 31. Let X be a Banach space and $\mathcal{B} \subseteq \mathcal{L}(X)$ be a bounded B.a. of projections. Show that if $P \in \langle\mathcal{B}\rangle_u^-$ is a projection, then actually $P \in \mathcal{B}$. ∎

Theorem III.2 shows, for any bounded B.a. of projections $\mathcal{B}$, that $\langle\mathcal{B}\rangle_u^-$ is isomorphic to $C(\Omega_\mathcal{B})$ both as a Banach space and as an algebra . That is, there exists a linear and multiplicative isomorphism $\Phi : C(\Omega_\mathcal{B}) \longrightarrow \langle\mathcal{B}\rangle_u^-$. Changing the situation somewhat, suppose that Λ is a compact Hausdorff space and $\Psi : C(\Lambda) \longrightarrow \mathcal{L}(X)$ is a linear and multiplicative map which is injective , continuous and has closed range. What can be said about the uniformly closed algebra $\Psi(C(\Lambda))$? Is it of the form $\langle\mathcal{B}\rangle_u^-$ for some bounded B.a. of projections $\mathcal{B}$? And, if so, how are the operators $\Psi(f) \in \mathcal{L}(X)$, for $f \in C(\Lambda)$, related to $\mathcal{B}$? The remainder of this chapter is devoted to a consideration of such questions.

Definition III.2.   Let Σ be an algebra of subsets of a set Ω and X be a Banach space. A map $P : \Sigma \longrightarrow \mathcal{L}(X)$ is called a *finitely additive spectral measure* if

   (i) $P(\Omega) = I$ and $P(\emptyset) = 0$,

   (ii) $P(\cup_{j=1}^n E_j) = \sum_{j=1}^n P(E_j)$, for all finite collections $\{E_j\}_{j=1}^n \subseteq \Sigma$ of pairwise disjoint sets, and

   (iii) $P(E \cap F) = P(E)P(F)$, for all $E, F \in \Sigma$, that is, P is *multiplicative*.

   Suppose that $\Gamma \subseteq X'$ is a vector space which distinguishes the points of X. If, *in addition* to (i)–(iii) above, Σ is a σ-algebra and P also satisfies the condition

   (iv) $E \mapsto \langle P(E)x, x' \rangle$, for $E \in \Sigma$, is a complex measure (i.e. is σ-additive) for every $x \in X$ and $x' \in \Gamma$,

then P is called a *spectral measure of class* Γ . The complex measure in (iv) is denoted by $\langle Px, x' \rangle$, for each $x \in X$ and $x' \in \Gamma$.   ■

Example 19. (a) Let $X = \ell^p$, $1 \le p < \infty$, and Σ denote the σ-algebra of all subsets of $\Omega = \mathbb{N}$. For each $E \in \Sigma$ define $P(E) \in \mathcal{L}(X)$ by

$$P(E) : x \mapsto (x_1 \chi_E(1), x_2 \chi_E(2), \dots), \qquad x = (x_1, x_2, \dots) \in X.$$

Then P is a spectral measure of class $\Gamma = X'$.

   (b) Let $X = \ell^\infty$ and (Ω, Σ) be as in (a). Define $P(E) \in \mathcal{L}(\ell^\infty)$ by the same formula as in (a). Then P is a spectral measure of class $\Gamma = \ell^1 \subset X'$ but *not* of class $\Gamma = X'$.   ■

   We require some further topologies on spaces of linear operators.

Definition III.3. Let X be a Banach space.

   (i) The *strong operator topology* on $\mathcal{L}(X)$ is defined by specifying a basic open neighbourhood of $T \in \mathcal{L}(X)$ by

$$\mathcal{N}_s(T) := \{R \in \mathcal{L}(X) : \ \|(R - T)x\| < \varepsilon, \quad x \in \mathcal{F}\},$$

where $\varepsilon > 0$ is arbitrary and $\mathcal{F}$ is any finite subset of X. Hence, a net $\{T_\alpha\} \subseteq \mathcal{L}(X)$ converges to $T \in \mathcal{L}(X)$ in the strong operator topology if and only if

$$(11) \qquad\qquad\qquad \lim_\alpha T_\alpha x = Tx, \qquad x \in X,$$

where the limit (11) exists in the norm topology of X. We also say that $\{T_\alpha\}$ converges strongly to T. The space $\mathcal{L}(X)$ equipped with the strong operator topology is denoted by $\mathcal{L}_s(X)$.

   (ii) The *weak operator topology* on $\mathcal{L}(X)$ is defined by specifying a basic open neighbourhood of $T \in \mathcal{L}(X)$ by

$$\mathcal{N}_w(T) := \{R \in \mathcal{L}(X) : \ |\langle (T - R)x, x' \rangle| < \varepsilon, \ x \in \mathcal{F}, \ x' \in \mathcal{H}\}$$

where $\varepsilon > 0$ is arbitrary, and $\mathcal{F} \subseteq X$ and $\mathcal{H} \subseteq X'$ are arbitrary finite sets. Hence, a net $\{T_\alpha\} \subseteq \mathcal{L}(X)$ converges in the weak operator topology to $T \in \mathcal{L}(X)$ if and only if

$$\lim_\alpha \langle T_\alpha x, x' \rangle = \langle Tx, x' \rangle, \qquad x \in X, \ x' \in X'.$$

The space $\mathcal{L}(X)$ equipped with the weak operator topology is denoted by $\mathcal{L}_w(X)$. ∎

Exercise 32. Let X be a Banach space.

(a) Suppose that $\{T_\alpha\} \subseteq \mathcal{L}(X)$ is a net of operators converging strongly to $T \in \mathcal{L}(X)$. Show that $\{T_\alpha\}$ also converges to T with respect to the weak operator topology.

(b) Let $X = \ell^2$ and $\xi \in X$ be any fixed non-zero vector. For each $n \in \mathbb{N}$, let $e_n \in X$ have 1 in the n-th co-ordinate and 0 elsewhere. Define $T_n : X \longrightarrow X$ by $T_n x = \langle x, \xi \rangle e_n$, for $x \in X$, where $\langle x, \xi \rangle = \sum_{n=1}^\infty x_n \xi_n$. Show that $T_n \in \mathcal{L}(X)$, for each $n \in \mathbb{N}$, and that the sequence $\{T_n\}_{n=1}^\infty$ converges to the zero operator in the weak operator topology, but $\{T_n\}$ does *not* converge strongly to the zero operator. ∎

Proposition III.1. *Let X be a Banach space.*

(a) *(Uniform Boundedness Principle) Let $\mathcal{A} \subseteq \mathcal{L}(X)$. Then the following statements are equivalent:*

(i) $\sup\{\|T\| : T \in \mathcal{A}\} < \infty$.

(ii) $\sup\{\|Tx\| : T \in \mathcal{A}\} < \infty$ *for each $x \in X$* .

(iii) $\sup\{|\langle Tx, x'\rangle| : T \in \mathcal{A}\} < \infty$ *for each $x \in X$ and $x' \in X'$.*

(b) *(Banach-Steinhaus theorem) Let $\{T_\alpha\} \subseteq \mathcal{L}(X)$ be a net such that $\sup_\alpha \|T_\alpha\| < \infty$ and $Tx := \lim_\alpha T_\alpha x$ exists in X, for each $x \in X$. Then $T \in \mathcal{L}(X)$.*

(c) *A linear functional on the vector space $\mathcal{L}(X)$ is continuous for the strong operator topology if and only if it is continuous for the weak operator topology.*

(d) *A convex subset of $\mathcal{L}(X)$ has the same closure for the weak operator topology as it does for the strong operator topology.*

For (a) we refer to [14; p.66] and for (b) we refer to [14; p.55]. Parts (c) and (d) can be found in [14; Chapter VI, §1], for example.

Definition III.4. A finitely additive spectral measure $P : \Sigma \longrightarrow \mathcal{L}(X)$ defined on a σ-algebra Σ is simply called a *spectral measure* if it is σ-additive with respect to the strong operator topology. That is, whenever $\{E_n\}_{n=1}^\infty \subseteq \Sigma$ is a decreasing sequence with $\cap_{n=1}^\infty E_n = \emptyset$, then $\lim_{n\to\infty} P(E_n) = 0$ in $\mathcal{L}_s(X)$. ∎

Lemma III.2. *Let X be a Banach space and $P : \Sigma \longrightarrow \mathcal{L}(X)$ be a finitely additive spectral measure defined on a σ-algebra Σ. Then P is a spectral measure if and only if it is a spectral measure of class $\Gamma = X'$, that is, if and only if $\langle Px, x'\rangle$ is a complex measure for each $x \in X$ and $x' \in X'$.*

Proof. Suppose that P is a spectral measure. Fix $x \in X$ and $x' \in X'$. Let $E_n \downarrow \emptyset$ in Σ. Then $\lim_{n\to\infty} P(E_n)x = 0$ in the norm of X. Hence, also $\langle Px, x'\rangle(E_n) \longrightarrow 0$ since $|\langle P(E_n)x, x'\rangle| \le \|P(E_n)x\| \cdot \|x'\|$. Accordingly, $\langle Px, x'\rangle$ is σ-additive.

Conversely, suppose that P is a spectral measure of class $\Gamma = X'$. Fix $x \in X$. Then $\langle Px, x'\rangle$ is σ-additive, for each $x' \in X'$, and hence the X-valued set function $Px : E \mapsto P(E)x$, for $E \in \Sigma$, is (norm) σ-additive; see Proposition I.1. Exercise 9(b) then implies that $P(E_n)x \longrightarrow 0$ in X whenever $E_n \downarrow \emptyset$ in Σ. ∎

Exercise 33. Let $\Omega = \mathbb{N}$ and Σ be the σ-algebra of all subsets of Ω. For each $E \in \Sigma$ define $P(E) \in \mathcal{L}(\ell^\infty)$ by $P(E)x = (x_1 \chi_E(1), x_2 \chi_E(2), \dots)$, for $x \in \ell^\infty$. Show that $P : \Sigma \longrightarrow \mathcal{L}(\ell^\infty)$

is a spectral measure of class $\Gamma_1 = \ell^1$, but P is *not* a spectral measure of class $\Gamma_2 = (\ell^\infty)'$. That is, verify the claims made in Example 19(b).   ∎

Definition III.5. Let X be a Banach space, $\Gamma \subseteq X'$ distinguish the points of X, and $P : \Sigma \longrightarrow \mathcal{L}(X)$ be a spectral measure of class Γ defined on a σ-algebra Σ. A Σ-measurable function $f : \Omega \longrightarrow \mathbb{C}$ is called *P-integrable* if

(i) $\int_\Omega |f|\, d|\langle Px, x'\rangle| < \infty$, that is, $f \in L^1(\langle Px, x'\rangle)$ for all $x \in X$, $x' \in \Gamma$, and

(ii)   for each $E \in \Sigma$ there exists an element of $\mathcal{L}(X)$, necessarily unique and denoted by $\int_E f\, dP$, such that

$$\langle (\int_E f\, dP)x, x'\rangle = \int_E f\, d\langle Px, x'\rangle, \qquad x \in X, \; x' \in \Gamma. \qquad \blacksquare$$

Definition III.6. A finitely additive spectral measure $P : \Sigma \longrightarrow \mathcal{L}(X)$ defined on an algebra of sets Σ is called *bounded* if

$$\|P(\Sigma)\| := \sup\{\|P(E)\| : \; E \in \Sigma\} < \infty. \qquad \blacksquare$$

Lemma III.3.   *Let X be a Banach space and $P : \Sigma \longrightarrow \mathcal{L}(X)$ be a spectral measure . Then P is necessarily bounded.*

Proof.   It follows from Lemma III.2 that $\langle Px, x'\rangle$ is σ-additive , for all $x \in X$, $x' \in X'$ and hence, by Theorem I.1, that

$$\sup_{E \in \Sigma} |\langle P(E)x, x'\rangle| < \infty, \qquad x \in X, \; x' \in X'.$$

The conclusion follows from the Uniform Boundedness Principle ; see Proposition III.1(a).∎

Exercise 34.[∗]   Let X be a Banach space and $P : \Sigma \longrightarrow \mathcal{L}(X)$ be a spectral measure . For each $E \in \Sigma$ let $Q(E) = P(E)' \in \mathcal{L}(X')$ be the dual operator of $P(E)$. Show that the function $Q : \Sigma \longrightarrow \mathcal{L}(X')$ so defined is a bounded spectral measure of class $\Gamma = X \subseteq X''$.∎

Exercise 35.   A Banach space X is said to contain a copy of ℓ^∞ if there exists a closed subspace Y of X and a bicontinuous isomorphism of ℓ^∞ onto Y. The following result can be found in [8; p.23].

Fact. *Let X be a Banach space which does not contain a copy of ℓ^∞ and let $\Gamma \subseteq X'$ be a subspace which distinguishes the points of X. Suppose that $\sum_{n=1}^\infty x_n$ is a (formal) series in X such that every subseries is Γ-convergent in the sense that for each $A \subseteq \mathbb{N}$ there is $x_A \in X$ satisfying*

$$\sum_{n \in A} \langle x_n, x'\rangle = \langle x_A, x'\rangle, \qquad x' \in \Gamma.$$

Then $\sum_{n=1}^\infty x_n$ is (norm) unconditionally convergent.

Let X be a Banach space *not* containing a copy of ℓ^∞ and $P : \Sigma \longrightarrow \mathcal{L}(X)$ be a spectral measure of class $\Gamma \subseteq X'$. Show that P is actually a spectral measure (i.e. of class X').   ∎

Proposition III.2.   *Let X be a Banach space and $P : \Sigma \longrightarrow \mathcal{L}(X)$ be a spectral measure of class $\Gamma \subseteq X'$. If P is bounded , then every bounded Σ-measurable function is P-integrable.*

Proof. Let $\psi \in B^\infty(\Sigma)$. Then certainly $\psi \in L^1(\langle Px, x'\rangle)$, for each $x \in X$, $x' \in \Gamma$. By (10) of Chapter I applied to $m := P$ (interpreted as a bounded finitely additive vector measure with values in the Banach space $\mathcal{L}_u(X)$) there exists, for each $E \in \Sigma$, a unique operator $\int_E \psi \, dP \in \mathcal{L}(X)$ satisfying

$$(12) \qquad \langle \int_E \psi \, dP, \xi \rangle = \int_E \psi \, d\langle P, \xi \rangle, \qquad \xi \in (\mathcal{L}_u(X))'.$$

Since $\xi(T) := \langle Tx, x'\rangle$, for $T \in \mathcal{L}(X)$, is an element of $(\mathcal{L}_u(X))'$ for each $x \in X$ and $x' \in \Gamma$, it is clear from (12) that (ii) of Definition III.5 is satisfied. Accordingly, ψ is P-integrable. ∎

We are now able to describe the nature of continuous linear and multiplicative homomorphisms from $C(\Lambda)$ into $\mathcal{L}_u(X)$, where Λ is a compact Hausdorff space, in terms of a B.a. of projections ; see [15; Chapter XVII, Theorem 2.4].

Theorem III.3. *Let X be a Banach space, Λ be a compact Hausdorff space and let $S : C(\Lambda) \longrightarrow \mathcal{L}_u(X)$ be a continuous linear map which is also a unital algebra homomorphism (i.e. $S(fg) = S(f)S(g)$, for $f, g \in C(\Lambda)$, and $S(\mathbb{1}) = I$). Then there exists a unique bounded spectral measure $R : Bo(\Lambda) \longrightarrow \mathcal{L}(X')$ of class $\Gamma = X \subseteq X''$ such that*

(i) *$\langle x, Rx' \rangle : Bo(\Lambda) \longrightarrow \mathbb{C}$ is regular for each $x \in X$ and $x' \in X'$,*

and

(ii) *$S(f)' = \int_\Lambda f \, dR$, $f \in C(\Lambda)$.*

Proof. By assumption $\|S\| := \sup\{\|S(f)\| : \|f\|_\infty \leq 1\} < \infty$.

Fix $x \in X$ and define $S_x : C(\Lambda) \longrightarrow X$ by $S_x(f) = S(f)x$, for $f \in C(\Lambda)$. By Theorem I.12 there exists a *unique* finitely additive measure $m_x : Bo(\Lambda) \longrightarrow X''$ such that $\langle x', m_x \rangle$ is a regular σ-*additive* measure for each $x' \in X'$ and satisfies

$$(13) \qquad \int_\Lambda f \, d\langle x', m_x \rangle = \langle S_x(f), x' \rangle = \langle S(f)x, x' \rangle, \qquad f \in C(\Lambda).$$

Let $\alpha_1, \alpha_2 \in \mathbb{C}$ and $x_1, x_2 \in X$. Then $S_{\alpha_1 x_1 + \alpha_2 x_2} = \alpha_1 S_{x_1} + \alpha_2 S_{x_2}$. Moreover, the mapping $\alpha_1 m_{x_1} + \alpha_2 m_{x_2} : Bo(\Lambda) \longrightarrow X''$ satisfies $\langle x', \alpha_1 m_{x_1} + \alpha_2 m_{x_2} \rangle = \alpha_1 \langle x', m_{x_1} \rangle + \alpha_2 \langle x', m_{x_2} \rangle$, is regular and σ-additive for each $x' \in X'$, and satisfies

$$\int_\Lambda f \, d\langle x', \alpha_1 m_{x_1} + \alpha_2 m_{x_2} \rangle = \langle S(f)(\alpha_1 x_1 + \alpha_2 x_2), x' \rangle, \qquad f \in C(\Lambda).$$

By *uniqueness* we deduce that

$$(14) \qquad \alpha_1 m_{x_1} + \alpha_2 m_{x_2} = m_{\alpha_1 x_1 + \alpha_2 x_2}.$$

Fix $E \in Bo(\Lambda)$. Then $x \mapsto \langle x', m_x(E) \rangle$, for $x \in X$, is linear (by (14)) and satisfies $|\langle x', m_x(E) \rangle| \leq \|\langle x', m_x \rangle\|$. Since $C(\Lambda)' = M(Bo(\Lambda))$, it follows from (13) that

$$\begin{aligned} \|\langle x', m_x \rangle\| &= \sup_{\|f\|_\infty \leq 1} \left| \int_\Lambda f \, d\langle x', m_x \rangle \right| = \\ &= \sup_{\|f\|_\infty \leq 1} |\langle S(f)x, x' \rangle| \leq \|S\| \cdot \|x\| \cdot \|x'\|, \end{aligned}$$

for *every* $x' \in X'$. This shows, for each fixed $x' \in X'$, that $x \mapsto \langle x', m_x(E) \rangle$, for $x \in X$, is a continuous linear functional on X. Hence, there is a unique vector $R(E)x' \in X'$ such that

$$(15) \qquad\qquad \langle x, R(E)x' \rangle = \langle x', m_x(E) \rangle.$$

Furthermore, we see from above that

$$\|R(E)x'\| := \sup_{\|x\| \le 1} |\langle x, R(E)x' \rangle| = \sup_{\|x\| \le 1} |\langle x', m_x(E) \rangle| \le \|S\| \cdot \|x'\|,$$

for all $x' \in X'$ and $E \in Bo(\Lambda)$. Moreover, $x' \mapsto R(E)x'$ is linear for each fixed $E \in Bo(\Lambda)$; this follows from (15) since, for each $\alpha_1, \alpha_2 \in \mathbb{C}$ and $x'_1, x'_2 \in X'$, we have that

$$
\begin{aligned}
\langle x, R(E)(\alpha_1 x'_1 + \alpha_2 x'_2) \rangle &= \langle \alpha_1 x'_1 + \alpha_2 x'_2, m_x(E) \rangle \\
&= \alpha_1 \langle x'_1, m_x(E) \rangle + \alpha_2 \langle x'_2, m_x(E) \rangle \\
&= \alpha_1 \langle x, R(E)x'_1 \rangle + \alpha_2 \langle x, R(E)x'_2 \rangle \\
&= \langle x, \alpha_1 R(E)x'_1 + \alpha_2 R(E)x'_2 \rangle.
\end{aligned}
$$

Since this is true for all $x \in X$, we conclude that

$$R(E)(\alpha_1 x'_1 + \alpha_2 x'_2) = \alpha_1 R(E)x'_1 + \alpha_2 R(E)x'_2.$$

So, $x' \mapsto R(E)x'$ is indeed linear for each $E \in Bo(\Lambda)$. Moreover, R is bounded since

$$\sup\{\|R(E)\| : E \in Bo(\Lambda)\} \le \|S\|.$$

Hence, $R : Bo(\Lambda) \longrightarrow \mathcal{L}(X')$ is a finitely additive measure with bounded range such that $\langle x, Rx' \rangle$ is a regular σ-additive measure, for each $x \in X$, $x' \in X'$, and satisfies

$$(16) \qquad\qquad \langle S(f)x, x' \rangle = \int_\Lambda f \, d\langle x, Rx' \rangle, \qquad f \in C(\Lambda).$$

We now show that R is a *finitely additive spectral measure* . Put $f = \mathbb{1}$ in (16) and use $S(\mathbb{1}) = I \in \mathcal{L}(X)$ gives

$$\langle x, x' \rangle = \langle x, R(\Lambda)x' \rangle, \qquad x \in X,\, x' \in X',$$

and so $R(\Lambda) = I \in \mathcal{L}(X')$. Put $f = 0$ gives $R(\emptyset) = 0$. The finite additivity of R implies that $R(E^c) = I - R(E)$, for $E \in Bo(\Lambda)$. It remains to check the multiplicativity of R.

Fix $g \in C(\Lambda)$ and $x \in X$, $x' \in X'$. Define a regular complex measure $\mu_{x,x'}$ by $\mu_{x,x'}(E) := \int_E g \, d\langle x, Rx' \rangle$, for $E \in Bo(\Lambda)$. Then, for *each* $f \in C(\Lambda) \subseteq L^1(\mu_{x,x'})$ we have (by (16) and the identity $S(fg) = S(f)S(g)$) that

$$
\begin{aligned}
\int_\Lambda f \, d\mu_{x,x'} &= \int_\Lambda fg \, d\langle x, Rx' \rangle = \langle S(f)S(g)x, x' \rangle = \\
&= \langle S(f)x, S(g)'x' \rangle = \int_\Lambda f \, d\langle x, RS(g)'x' \rangle.
\end{aligned}
$$

Since $\mu_{x,x'}$ and $\langle x, R(\cdot)S(g)'x' \rangle$ are both regular complex measures it follows that $\mu_{x,x'} = \langle x, R(\cdot)S(g)'x' \rangle$ as an equality of measures, that is,

$$(17) \qquad \int_E g\, d\langle x, Rx' \rangle = \langle x, R(E)S(g)'x' \rangle, \qquad g \in C(\Lambda),\ E \in Bo(\Lambda).$$

Suppose we know that $S(g)'R(E) = R(E)S(g)'$ for all $E \in Bo(\Lambda)$ and $g \in C(\Lambda)$. Then it follows from (17), for E *fixed* now, that

$$\int_E g\, d\langle x, Rx' \rangle = \langle x, S(g)'R(E)x' \rangle = \langle S(g)x, R(E)x' \rangle = \int_\Lambda g\, d\langle x, Rz' \rangle,$$

where $z' = R(E)x'$. That is,

$$\int_\Lambda g\, d\langle x, Rx' \rangle \big|_E = \int_\Lambda g\, d\langle x, Rz' \rangle, \qquad g \in C(\Lambda),$$

where $\langle x, Rx' \rangle\big|_E$ denotes the restriction of the measure $\langle x, Rx' \rangle$ to E. By uniqueness of regular measures we deduce that $\langle x, Rx' \rangle\big|_E = \langle x, Rz' \rangle$ as an equality of measures, that is,

$$\langle x, R(E \cap F)x' \rangle = \langle x, R(F)z' \rangle = \langle x, R(F)R(E)x' \rangle, \qquad F \in Bo(\Lambda).$$

By also fixing F we conclude, since $x \in X$ and $x' \in X'$ are arbitrary, that

$$R(E \cap F) = R(F)R(E).$$

So, it remains to verify that $S(g)'R(E) = R(E)S(g)'$, for all $g \in C(\Lambda)$ and $E \in Bo(\Lambda)$. Accordingly, fix g and E. Choose $Bo(\Lambda)$ -simple functions $\{g_n\}_{n=1}^\infty$ such that $\|g_n - g\|_\infty \longrightarrow 0$. We have seen before that

$$\Big\| \int_\Lambda g_n\, dR \Big\| \leq 4 \cdot \|S\| \cdot \|g_n\|_\infty, \qquad n \in \mathbb{N},$$

and so there exists $R(g) \in \mathcal{L}(X')$ with $\int_\Lambda g_n\, dR \longrightarrow R(g)$ in $\mathcal{L}_u(X')$. Since $R(E)$ commutes with each operator $\int_\Lambda g_n\, dR$, for $n \in \mathbb{N}$, it is clear that $R(E)R(g) = R(g)R(E)$. So, it suffices to show that $R(g) = S(g)'$. By the dominated convergence theorem applied to $\langle x, Rx' \rangle$ we see (c.f. (16)) that

$$\langle x, S(g)'x' \rangle = \langle S(g)x, x' \rangle = \int_\Lambda g\, d\langle x, Rx' \rangle = \lim_{n \to \infty} \int_\Lambda g_n\, d\langle x, Rx' \rangle.$$

But, $\int_\Lambda g_n\, d\langle x, Rx' \rangle = \langle x, (\int_\Lambda g_n\, dR)x' \rangle \longrightarrow \langle x, R(g)x' \rangle$, as $n \to \infty$, since $\int_\Lambda g_n\, dR \longrightarrow R(g)$ in $\mathcal{L}_u(X')$. Hence, we deduce that

$$\langle x, S(g)'x' \rangle = \langle x, R(g)x' \rangle, \qquad x \in X,\ x' \in X',$$

which implies the desired equality $S(g)' = R(g)$. ∎

An unpleasant feature of Theorem III.3 is that the B.a. of projections $\{R(E) : E \in Bo(\Lambda)\}$ exists in $\mathcal{L}(X')$ rather than in $\mathcal{L}(X)$. If the Banach space X in Theorem III.3 does not contain an isomorphic copy of c_0, then more can be said. Indeed, the bounded linear maps $S_x : C(\Lambda) \longrightarrow X$ given by $S_x : f \mapsto S(f)x$ are then necessarily *weakly compact* ; see Theorem I.14. Then applying the vector-valued Riesz representation theorem (see Theorem I.13) in place of Theorem I.12 it is possible to choose for each $x \in X$ a *unique* σ-additive vector measure $m_x : Bo(\Lambda) \longrightarrow X$ (rather than X''-valued) which is regular and satisfies

$$S(f)x = \int_\Lambda f\, dm_x, \qquad f \in C(\Lambda).$$

An analogues argument as in the proof of Theorem III.3 then establishes the following result.

Theorem III.4.   *Let Λ be a compact Hausdorff space, X be a Banach space not containing an isomorphic copy of c_0 and $S : C(\Lambda) \longrightarrow \mathcal{L}_u(X)$ be a continuous linear map which is also a unital homomorphism . Then there exists a unique spectral measure $P : Bo(\Lambda) \longrightarrow \mathcal{L}(X)$ such that*

(i) *each vector measure $Px : Bo(\Lambda) \longrightarrow X$, for $x \in X$, given by $E \mapsto P(E)x$ is regular* and

(ii) $S(f) = \int_\Lambda f\, dP, \qquad f \in C(\Lambda).$

Let $\mathcal{B}$ be a *bounded* B.a. of projections in $\mathcal{L}(X)$ and let $\langle \mathcal{B} \rangle_u^-$ be the uniformly closed algebra generated by $\mathcal{B}$ in $\mathcal{L}_u(X)$. By Theorem III.2 there exists a *finitely additive spectral measure* $Q : Co(\Omega_\mathcal{B}) \longrightarrow \mathcal{L}(X)$, where $\Omega_\mathcal{B}$ is the Stone space of $\mathcal{B}$, such that the map $S : C(\Omega_\mathcal{B}) \longrightarrow \mathcal{L}(X)$ given by

(18)
$$S(f) = \int_{\Omega_\mathcal{B}} f\, dQ, \qquad f \in C(\Omega_\mathcal{B}),$$

is a Banach algebra isomorphism of $C(\Omega_\mathcal{B})$ onto $\langle \mathcal{B} \rangle_u^-$. Recall that $\int_\Omega f\, dQ$ is defined by a continuous extension process from the dense subalgebra $\mathrm{sim}(Co(\Omega_\mathcal{B}))$ to all of $C(\Omega_\mathcal{B})$. We note that Q is only defined on the *algebra of sets* $Co(\Omega_\mathcal{B})$.

By Theorem III.3 there exists a unique regular, bounded spectral measure $R : Bo(\Omega_\mathcal{B}) \longrightarrow \mathcal{L}(X')$ of *class* $\Gamma = X \subseteq X''$ (so some σ-additivity is present) such that

(19)
$$S(f)' = \int_{\Omega_\mathcal{B}} f\, dR, \qquad f \in C(\Omega_\mathcal{B}),$$

where S is given by (18). The first point is that R is defined on the σ-*algebra* $Bo(\Omega_\mathcal{B})$ and takes its values in $\mathcal{L}(X')$, with X' the *dual space* of X. If $E \in Co(\Omega_\mathcal{B})$, then $\chi_E \in C(\Omega_\mathcal{B})$ and so we see from (18) and (19) that $R(E) = Q(E)'$. So, every projection in the B.a. $\{R(E) : E \in Co(\Omega_\mathcal{B})\} \subseteq \mathcal{L}(X')$ is the dual operator of some projection from the B.a. $\mathcal{B} = \{Q(E) : E \in Co(\Omega_\mathcal{B})\} \subseteq \mathcal{L}(X)$.

Fix $x \in X$ and $x' \in X'$. Then the finitely additive measure $\langle Qx, x' \rangle$ is actually σ-additive *on the algebra* $Co(\Omega_\mathcal{B})$. Indeed, suppose that $\{E_n\}_{n=1}^\infty \subseteq Co(\Omega_\mathcal{B})$ is a sequence of pairwise disjoint sets such that $E := \cup_{n=1}^\infty E_n$ belongs to $Co(\Omega_\mathcal{B})$. Since $E \in Co(\Omega_\mathcal{B})$, it is closed hence compact. Accordingly, $\{E_n\}_{n=1}^{\infty}$ is an open cover of the compact set E and so there exist finitely many sets $\{E_{n_j}\}_{j=1}^{k}$ such that $E = \cup_{j=1}^{k} E_{n_j}$. Moreover, the disjointness property implies that $E_n = \emptyset$ for $n \notin \{n_j\}_{j=1}^{k}$. Then the finite additivity of Q yields

$$\langle Q(\cup_{n=1}^{\infty} E_n)x, x' \rangle = \sum_{n=1}^{\infty} \langle Q(E_n)x, x' \rangle.$$

This establishes the σ-additivity of Q on the algebra of sets $Co(\Omega_B)$.

An examination of Definition I.1 (as already noted) shows that the variation $|\nu|$ of any *bounded finitely additive* function $\nu : S \longrightarrow \mathbb{C}$ defined on an *algebra of sets* S is just as well defined as for a complex measure with domain a σ-algebra of sets.

Definition III.7. A bounded finitely additive measure $\nu : S \longrightarrow \mathbb{C}$ defined on an *algebra of sets* S of some topological space Λ is said to be *regular* if for each $E \in S$ and each $\varepsilon > 0$ there is a set $K \in S$ with $\overline{K} \subseteq E$ and a set $U \in S$ with $E \subseteq U^\circ$ such that $|\nu|(U \backslash K) < \varepsilon$. This is equivalent to $|\nu(F)| < \frac{\varepsilon}{4}$ for all $F \in S$ with $F \subseteq U \backslash K$. $\blacksquare$

The following classical extension theorem for measures is due to A.D. Alexandroff ; see [14; p.138], for example.

Theorem III.5. *Let $\nu : S \longrightarrow \mathbb{C}$ be a bounded, regular, finitely additive measure defined on an algebra of sets S of some compact topological Hausdorff space Λ. Then ν is σ-additive on S and ν has a unique extension to a regular, σ-additive measure on $\sigma(S)$, the σ-algebra generated by S.*

Exercise 36. Let $\mathcal{A}$ be an *algebra of subsets* of a set Λ and let $\Sigma = \sigma(\mathcal{A})$ be the σ-algebra generated by $\mathcal{A}$. Let $\mu : \Sigma \longrightarrow \mathbb{C}$ and $\nu : \Sigma \longrightarrow \mathbb{C}$ be complex measures such that $\mu(E) = \nu(E)$ for every $E \in \mathcal{A}$. Show that $\mu(E) = \nu(E)$ for every $E \in \Sigma$. $\blacksquare$

After this short digression let us return to the situation of a *bounded* B.a. of projections $\mathcal{B} \subseteq \mathcal{L}(X)$ and its Stone map $Q : Co(\Omega_B) \longrightarrow \mathcal{B}$, where Ω_B is the Stone space of $\mathcal{B}$. Fix $x \in X$ and $x' \in X'$. It was noted above that $\langle Qx, x' \rangle$ is σ-additive on the *algebra of sets* $Co(\Omega_B)$. Moreover, we have seen that

$$\sup\{|\langle Q(E)x, x' \rangle| : E \in Co(\Omega_B)\} \leq 4\|\mathcal{B}\| \cdot \|x\| \cdot \|x'\|, \qquad x \in X, \, x' \in X',$$

showing that $\langle Qx, x' \rangle$ is a *bounded* measure. It is a routine observation to note that Definition III.7 is fulfilled with $S = Co(\Omega_B)$ and so $\langle Qx, x' \rangle$ is regular on $Co(\Omega_B)$. Proposition II.5 and Theorem III.5 imply that $\langle Qx, x' \rangle$ has a unique, *regular* extension $\mu_{x,x'} : Ba(\Omega_B) \longrightarrow \mathbb{C}$ which is also σ-additive. By Theorem III.3 there also exists a regular, σ-additive measure $\langle x, Rx' \rangle : Bo(\Omega_B) \longrightarrow \mathbb{C}$ which coincides with $\langle Qx, x' \rangle$ on $Co(\Omega_B)$. Then necessarily $\mu_{x,x'} = \langle x, Rx' \rangle$ on the σ-algebra $Ba(\Omega_B) \subseteq Bo(\Omega_B)$; see Exercise 36. Actually, $\langle x, Rx' \rangle$ is the *unique* regular Borel extension of the Baire measure $\mu_{x,x'}$; see [37; p.314].

Part of our aim in the next two chapters will be to investigate more closely certain properties of the B.a. $\mathcal{B}$ which ensure that the extension of the scalar measures $\langle Qx, x' \rangle$ from $Co(\Omega_B)$ to $Ba(\Omega_B)$ or $Bo(\Omega_B)$ is effected by *members of $\mathcal{B}$ itself*. That is, for each E from $Ba(\Omega_B)$ or $Bo(\Omega_B)$ we require the existence of $Q_E \in \mathcal{B}$ which satisfies

$$\mu_{x,x'}(E) = \langle Q_E x, x' \rangle = \langle x, R(E)x' \rangle, \qquad x \in X, \, x' \in X'.$$

Equivalently, every projection from $\{R(E) : E \in Ba(\Omega_{\mathcal{B}})\}$ or $\{R(E) : E \in Bo(\Omega_{\mathcal{B}})\}$, rather than just those from $\{R(E) : E \in Co(\Omega_{\mathcal{B}})\}$, should be the dual operator of some projection from $\mathcal{B}$. The appropriate conditions on $\mathcal{B}$ which ensure that this is the case will be formulated in the following chapters.

Chapter IV

Ranges of spectral measures and Boolean algebras of projections

At the end of Chapter II we saw that the Stone map $Q : Co(\Omega_B) \longrightarrow B$ can be extended to a σ-homomorphism $\overline{Q}$ (resp. $\widehat{Q}$) on $Ba(\Omega_B)$ (resp. $Bo(\Omega_B)$) provided that the B.a. B is abstractly σ-complete (resp. abstractly complete). In this generality not much more can be said. However, the more specialized case of a B.a. of *projections* B acting in a Banach space X has the additional feature that it is part of a vector space equipped with a *topology* , namely $\mathcal{L}_s(X)$ or $\mathcal{L}_w(X)$. Since both $Ba(\Omega_B)$ and $Bo(\Omega_B)$ are σ-*algebras of sets*, it then has a meaning to ask whether $\overline{Q}$ and/or $\widehat{Q}$ are σ-additive, that is, whether they are *spectral measures*? The abstract σ-completeness or abstract completeness of B by itself no longer suffices in this case; to see this combine Example 20(a) with Theorem IV.1 below. It turns out that this question has a precise (and positive) answer in terms of some very natural properties of B which intimately relate certain order properties of directed systems in B with some topological requirements. Such conditions, already known quite some time ago for B.a. 's of *selfadjoint* projections in a Hilbert space (see [18], [25], for example), were extended to the Banach space setting by W. Bade [1,2]. The purpose of this chapter is to make a detailed study of the class of those B.a. 's of projections which can be represented as the range of a σ-*additive* spectral measure defined on a σ-algebra of sets. For such B.a. 's the powerful methods of vector measures and integration can be invoked; this will be exploited in the following chapter.

Let us begin with various notions which connect the order properties of B with some topological requirements.

Definition IV.1. Let X be a Banach space and $B \subseteq \mathcal{L}(X)$ be a B.a. of projections.

(i) B is called *Bade complete* (resp. *Bade σ-complete*) if it is abstractly complete (resp. abstractly σ-complete) in the sense of Definition II.4 and if

$$(\wedge_\alpha B_\alpha)X = \cap_\alpha B_\alpha X \quad \text{and} \quad (\vee_\alpha B_\alpha)X = \overline{sp\{\cup_\alpha B_\alpha X\}}$$

whenever $\{B_\alpha\}$ is a family (resp. countable family) of elements from B.

(ii) $\mathcal{B}$ has the *monotone* (resp. σ-*monotone*) *property* if $\lim_\alpha B_\alpha$ exists for the weak operator topology and belongs to $\mathcal{B}$ whenever $\{B_\alpha\} \subseteq \mathcal{B}$ is a monotone net (resp. sequence) with respect to the partial order of $\mathcal{B}$.

(iii) $\mathcal{B}$ has the *ordered* (resp. σ-*ordered*) *convergence property* if $\lim_\alpha B_\alpha$ exists for the strong operator topology and belongs to $\mathcal{B}$ whenever $\{B_\alpha\} \subseteq \mathcal{B}$ is a monotone net (resp. sequence) with respect to the partial order of $\mathcal{B}$.   ■

Definition IV.1(i), without the term "Bade", was introduced by W. Bade in [1].

Example 20.   (a) Let $X = \ell^\infty$ and, for each $E \in 2^\mathbb{N}$, let $P(E) \in \mathcal{L}(X)$ be defined by

$$P(E)x = (x_1\chi_E(1), x_2\chi_E(2), \dots), \qquad x = (x_1, x_2, \dots) \in \ell^\infty.$$

Then $\mathcal{B} := \{P(E) : \ E \in 2^\mathbb{N}\}$ is an abstractly complete B.a. of projections which *fails* to be Bade σ-complete. Indeed, let $E_n = \{n\}$, for $n \in \mathbb{N}$, in which case $\vee_n P(E_n) = I$. But, the linear hull $\mathrm{sp}\{\cup_{n=1}^\infty P(E_n)X\}$ is the subspace of ℓ^∞ consisting of all elements with only finitely many non-zero co-ordinates. Hence, $\overline{\mathrm{sp}\{\cup_{n=1}^\infty P(E_n)X\}} = c_0$ whereas $(\vee_n P(E_n))X = \ell^\infty$.

(b) Let $X = \ell^2([0,1])$, a *nonseparable* Hilbert space. Let $\Sigma = Bo([0,1])$ and, for each $E \in \Sigma$, let $P(E) \in \mathcal{L}(X)$ be the projection $P(E) : \phi \mapsto \chi_E\phi$ (pointwise multiplication), for $\phi \in X$. Then $\mathcal{B} := P(\Sigma)$ is Bade σ-complete. Let $A \subseteq [0,1]$ be a set which is *not* a Borel set. If $\mathcal{F}(A)$ is the family of all finite subsets of A, then $\{P(E) : E \in \mathcal{F}(A)\}$ is a family of projections belonging to $\mathcal{B}$ which has no least upper bound in $\mathcal{B}$. Hence, $\mathcal{B}$ is not abstractly complete and so also fails to be Bade complete.

(c) Let $X = \ell^2$ and Σ be the B.a. of all subsets of $\mathbb{N}$ which are finite or the complement of a finite set. For each $E \in \Sigma$ define a projection $B_E \in \mathcal{L}(X)$ by

$$B_E x = (x_1\chi_E(1), x_2\chi_E(2), \dots), \qquad x = (x_1, x_2, \dots) \in \ell^2.$$

Then $\mathcal{B} := \{B_E : E \in \Sigma\}$ is a bounded B.a. of projections which is not abstractly σ-complete. Indeed, if $E_n = \{2, 4, \dots, 2n\}$, for $n \in \mathbb{N}$, then $\{B_{E_n}\}_{n=1}^\infty$ has no least upper bound in $\mathcal{B}$.   ■

Let X be a Banach space and $P : \Sigma \longrightarrow \mathcal{L}(X)$ be a spectral measure defined on a σ-algebra of sets Σ of some set $\Omega \neq \emptyset$. A set $E \in \Sigma$ is called P-*null* if $P(E) = 0$. By the multiplicativity of P (c.f. Definition III.2) this is equivalent to $P(F) = 0$ for all $F \in \Sigma$ with $F \subseteq E$. Two sets $E, F \in \Sigma$ are called P-*equivalent* if their symmetric difference $E\triangle F$ is P-null. This is an equivalence relation in Σ; the equivalence class of $E \in \Sigma$ is denoted by $[E]$ and the family of all equivalence classes $\{[E] : \ E \in \Sigma\}$ is denoted by $\Sigma(P)$. It is routine to check that the operations

$$[E] \wedge [F] := [E \wedge F] \quad \text{and} \quad [E] \vee [F] := [E \vee F] \quad \text{and} \quad [E]' := [E'],$$

with $E \wedge F = E \cap F$ and $E \vee F = E \cup F$ and $E' = E^c$ being the B.a. operations in Σ, are well-defined in $\Sigma(P)$ and turn $\Sigma(P)$ into a B.a. Moreover, if $\mathcal{B} := P(\Sigma)$, then the induced map $\widetilde{P} : \Sigma(P) \longrightarrow \mathcal{B}$ given by $\widetilde{P}([E]) := P(E)$ is a *B.a. isomorphism* of $\Sigma(P)$ onto $\mathcal{B}$. Note that the zero element $[\emptyset]$ of $\Sigma(P)$ consists of all the P-null sets in Σ. Part (a) of the following exercise shows that $\widetilde{P}$ is well defined.

Exercise 37. Let X be a Banach space and $P : \Sigma \longrightarrow \mathcal{L}(X)$ be a spectral measure.

(a) If $E, F \in \Sigma$ are P-equivalent (i.e. $P(E \triangle F) = 0$) show that $P(E) = P(F)$.

(b) Let $\{E_j\}_{j=1}^n \subseteq \Sigma$ be P-null sets. Show that $\cup_{j=1}^n E_j$ is also P-null. ∎

Recall that a function $\rho : Y \times Y \longrightarrow [0, \infty)$, where Y is any non-empty set, is called a *pseudometric* if it satisfies $\rho(u, v) = \rho(v, u)$, for all $u, v \in Y$, and $\rho(u, v) \leq \rho(u, z) + \rho(z, v)$ for all $u, v, z \in Y$; see [24; Chapter 15].

Let $P : \Sigma \longrightarrow \mathcal{L}(X)$ be a spectral measure . For each $x \in X$, define a pseudometric d_x on $\Sigma(P)$ by

$$d_x([E], [F]) := \|Px\|(E \triangle F), \qquad [E], [F] \in \Sigma(P),$$

where $\|Px\|$ is the semivariation of the *vector measure* $Px : \Sigma \longrightarrow X$ given by $E \mapsto P(E)x$, for $E \in \Sigma$. The triangle inequality for d_x follows from $E \triangle F \subseteq (E \triangle G) \cup (G \triangle F)$, valid for any $E, F, G \in \Sigma$, together with the subadditivity of semivariation; see Lemma I.1(b). The topology and *uniform structure* on $\Sigma(P)$ specified by the family of pseudometrics $\{d_x : x \in X\}$ (see [24; Chapter 15] for the definition of uniform structure) is denoted by $\tau_s(P)$.

Exercise 38. Let $P : \Sigma \longrightarrow \mathcal{L}(X)$ be a spectral measure . Show that the family of pseudometrics $\{d_x : x \in X\}$ has the Hausdorff property, namely that $[E] = [F]$ whenever $[E], [F] \in \Sigma(P)$ satisfy $d_x([E], [F]) = 0$ for all $x \in X$. ∎

Definition IV.2. A spectral measure $P : \Sigma \longrightarrow \mathcal{L}(X)$ is called a *closed spectral measure* if $(\Sigma(P), \tau_s(P))$ is a *complete* uniform space. ∎

The previous definition is a special case (for spectral measures) of the notion of a closed measure for arbitrary *vector* measures taking values in any locally convex Hausdorff space; see [27; Chapter IV]. For the definition and basic theory of locally convex Hausdorff spaces we refer to [14; Chapter V].

The main aim of this chapter is to establish the connection between the various notions of completeness for a B.a. of projections (as given in Definition IV.1) and spectral measures.

Theorem IV.1. *Let X be a Banach space and $\mathcal{B} \subseteq \mathcal{L}(X)$ be a B.a. of projections. Then the following statements are equivalent.*

(a) $\mathcal{B}$ *has the σ-monotone (resp. monotone) property.*

(b) $\mathcal{B}$ *has the σ-ordered (resp. ordered) convergence property.*

(c) $\mathcal{B}$ *is Bade σ-complete (resp. Bade complete).*

(d) $\mathcal{B}$ *coincides with the range of some spectral (resp. closed spectral) measure.*

The proof of Theorem IV.1 will require a series of lemmas and propositions. The extension of the Stone map Q from the algebra of sets $Co(\Omega_\mathcal{B})$ to the σ-algebras $Ba(\Omega_\mathcal{B})$ and $Bo(\Omega_\mathcal{B})$, as given by Theorem II.4 and Theorem II.5, will play a decisive role. We begin with a useful technical result.

Lemma IV.1. *Let X be a Banach space and $\mathcal{B} \subseteq \mathcal{L}(X)$ be a B.a. of projections.*

(a) *Let $\mathcal{B}$ have the monotone (resp. σ-monotone) property. Then*

(i) $\mathcal{B}$ *is abstractly complete (resp. abstractly σ-complete) , and*

(ii) *whenever $\{B_\alpha\} \subseteq \mathcal{B}$ is an increasing net (resp. sequence), then $\lim_\alpha B_\alpha = \vee_\alpha B_\alpha$, whereas if $\{B_\alpha\}$ is a decreasing net (resp. sequence), then $\lim_\alpha B_\alpha = \wedge_\alpha B_\alpha$, where the limits* *exist in $\mathcal{L}_w(X)$.*

(b) *Let $\mathcal{B}$ have the ordered (resp. σ-ordered) convergence property. Then*

(i) *$\mathcal{B}$ is abstractly complete (resp. abstractly σ-complete) , and*

(ii) *whenever $\{B_\alpha\}$ is an increasing net (resp. sequence), then $\lim_\alpha B_\alpha = \vee_\alpha B_\alpha$, whereas if $\{B_\alpha\}$ is a decreasing net (resp. sequence), then $\lim_\alpha B_\alpha = \wedge_\alpha B_\alpha$, where the limits exist in $\mathcal{L}_s(X)$.*

Proof. (a) (ii) Suppose that $\mathcal{B}$ has the monotone property. Let $\{B_\alpha\}_{\alpha \in A} \subseteq \mathcal{B}$ be an increasing net. By definition of the monotone property $B = \lim_\alpha B_\alpha$ exists in $\mathcal{L}_w(X)$ and $B \in \mathcal{B}$. If $\alpha \in A$, then $B_\alpha B_\gamma = B_\alpha$ for all $\gamma \geq \alpha$ and so

$$\langle BB_\alpha x, x' \rangle = \lim_\gamma \langle B_\gamma B_\alpha x, x' \rangle = \langle B_\alpha x, x' \rangle, \qquad x \in X, \, x' \in X',$$

which implies that $BB_\alpha = B_\alpha$. That is, $B_\alpha \leq B$ for all $\alpha \in A$. Suppose that $D \in \mathcal{B}$ satisfies $B_\alpha \leq D$ for all $\alpha \in A$. Then $B_\alpha D = B_\alpha$ for all $\alpha \in A$ and so, for each $x \in X$ and $x' \in X'$, we have

$$\langle BDx, x' \rangle = \lim_\alpha \langle B_\alpha Dx, x' \rangle = \lim_\alpha \langle B_\alpha x, x' \rangle = \langle Bx, x' \rangle,$$

which implies that $BD = B$, i.e. $B \leq D$. This establishes that $\vee_\alpha B_\alpha$ exists in $\mathcal{B}$ and equals the weak operator limit $\lim_\alpha B_\alpha$. A similar argument establishes that $\wedge_\alpha B_\alpha$ exists and equals the weak operator limit $\lim_\alpha B_\alpha$ whenever $\{B_\alpha\} \subseteq \mathcal{B}$ is a decreasing net.

(i) To show that $\mathcal{B}$ is abstractly complete let $\mathcal{A}$ be an arbitrary subset of $\mathcal{B}$. Let $\mathcal{F}(\mathcal{A})$ be the collection of all finite subsets of $\mathcal{A}$ and, for $F \in \mathcal{F}(\mathcal{A})$, define $B_F = \vee\{B : B \in F\}$. Then $\{B_F\}_{F \in \mathcal{F}(\mathcal{A})}$ is an increasing net in $\mathcal{B}$ if we turn $\mathcal{F}(\mathcal{A})$ into a *directed set* (see [13; p.26] for the definition) by defining $F \leq G$ in $\mathcal{F}(\mathcal{A})$ whenever $F \subseteq G$. Moreover, $\vee\{B : B \in \mathcal{A}\}$ exists in $\mathcal{B}$ if and only if $\vee_{F \in \mathcal{F}(\mathcal{A})} B_F$ exists in $\mathcal{B}$, in which case the two suprema are equal (see Exercise 39 below). But, we just proved above that $\vee_{F \in \mathcal{F}(\mathcal{A})} B_F$ does exist in $\mathcal{B}$ (and equals the weak operator limit $\lim_F B_F$) and hence, $\vee\{B : B \in \mathcal{A}\}$ exists in $\mathcal{B}$. A similar argument shows that $\wedge\{B : B \in \mathcal{A}\}$ exists in $\mathcal{B}$. Accordingly, $\mathcal{B}$ is abstractly complete.

The case when $\mathcal{B}$ is σ-monotone can be established via a similar argument.

(b) This follows from (a) since the convergence of a net in $\mathcal{L}_s(X)$ implies it also converges in $\mathcal{L}_w(X)$ and to the same limit. ∎

Exercise 39. (a) Let X be a Banach space and $\mathcal{B} \subseteq \mathcal{L}(X)$ be a B.a. of projections with the monotone property. Show that whenever $\{B_\alpha\} \subseteq \mathcal{B}$ is a decreasing net, then $\wedge_\alpha B_\alpha$ exists in $\mathcal{B}$ and equals $\lim_\alpha B_\alpha$ (in $\mathcal{L}_w(X)$).

(b) Let $\mathcal{B} \subseteq \mathcal{L}(X)$ be *any* B.a. of projections and let $\mathcal{A}$ be an arbitrary subset of $\mathcal{B}$. For each finite set $F \subseteq \mathcal{A}$ let $B_F = \vee\{B : B \in F\}$. Given finite subsets F_1 and F_2 of $\mathcal{A}$ define $F_1 \leq F_2$ if $F_1 \subseteq F_2$. Show that the collection $\mathcal{F}(\mathcal{A})$ of all finite subsets of $\mathcal{A}$ is a directed set with respect to $\leq$ and that the net $\{B_F\}_{F \in \mathcal{F}(\mathcal{A})}$ is increasing in $\mathcal{B}$. Moreover, show that $\vee\{B : B \in \mathcal{A}\}$ exists in $\mathcal{B}$ if and only if $\vee_{F \in \mathcal{F}(\mathcal{A})} B_F$ exists in $\mathcal{B}$ and that, in this case, the two suprema are equal. ∎

The next result is crucial for the proof of Theorem IV.1. It connects the σ-monotone property of $\mathcal{B}$ with the σ-additivity of the *extended* Stone maps $\overline{Q}$ and $\widehat{Q}$ (of Q) to $Ba(\Omega_\mathcal{B})$

and $Bo(\Omega_B)$, respectively.

Lemma IV.2. *Let X be a Banach space and $B \subseteq L(X)$ be a B.a. of projections. Let Ω_B be the Stone space of B and $Q : Co(\Omega_B) \longrightarrow B$ be the Stone map.*

(a) *There exists a unique spectral measure $P : Ba(\Omega_B) \longrightarrow L(X)$ having the properties that $B = P(Ba(\Omega_B))$ and P coincides with Q on $Co(\Omega_B)$, if and only if, B has the σ-monotone property .*

(b) *Suppose that the B.a. B is abstractly complete . Then there exists a spectral measure $P : Bo(\Omega_B) \longrightarrow L(X)$ having the properties that $B = P(Bo(\Omega_B))$ and P coincides with Q on $Co(\Omega_B)$, if and only if, B has the σ-monotone property .*

Proof. (a) Suppose there exists a spectral measure $P : Ba(\Omega_B) \longrightarrow L_s(X)$ such that $P(Ba(\Omega_B)) = B$ and P coincides with Q on $Co(\Omega_B)$. Let $\{B_n\}_{n=1}^\infty$ be an increasing sequence in B in which case $B_n = P(E_n)$ for some sets $E_n \in Ba(\Omega_B)$. Then

$$P(E_n \backslash E_{n+k}) = P(E_n) - P(E_n \cap E_{n+k}) = P(E_n) - P(E_n)P(E_{n+k}) = B_n - B_n B_{n+k} = 0,$$

for all n and k, since $B_n \le B_{n+k}$ means that $B_n B_{n+k} = B_n$. Then the sets $A_n = \cup_{j=1}^n E_j$ are increasing in Ω_B and

$$P(A_n) = P([\cup_{j=1}^{n-1}E_j]\backslash E_n) + P(E_n) = P(\cup_{j=1}^{n-1}(E_j\backslash E_n)) + B_n = B_n, \qquad n \in \mathbb{N},$$

since if $P(F_j) = 0$, for $1 \le j \le r$, then also $P(\cup_{j=1}^r F_j) = 0$; see Exercise 37(b). Since P is σ-additive in $L_s(X)$, we have that

$$\lim_{n\to\infty} B_n = \lim_{n\to\infty} P(A_n) = P(\cup_{k=1}^\infty A_k),$$

where $P(\cup_{k=1}^\infty A_k) \in P(Ba(\Omega_B)) = B$ and the limit exists in $L_s(X)$. If $\{B_n\} \subseteq B$ is a decreasing sequence we consider the increasing sequence $\{I - B_n\}_{n=1}^\infty \subseteq B$ to deduce again that $\lim_{n\to\infty} B_n$ exists in $L_s(X)$ and belongs to B. Since the existence of a limit in $L_s(X)$ implies its existence in $L_w(X)$ we conclude that B has the σ-monotone property.

Conversely, suppose that B has the σ-monotone property. By Lemma IV.1 we know that B is then abstractly σ-complete . Then Theorem II.4 guarantees that there exists a *unique* σ-homomorphism P from $Ba(\Omega_B)$ onto B such that P extends Q. Let $\{E_n\}_{n=1}^\infty \subseteq Ba(\Omega_B)$ decrease to $\emptyset$. By Lemma IV.1(a), since $\{P(E_n)\}_{n=1}^\infty$ is decreasing in B, it follows that

$$\lim_{n\to\infty} P(E_n) = \wedge_{n=1}^\infty P(E_n) = P(\cap_{n=1}^\infty E_n) = 0,$$

where the limit exists in $L_w(X)$. So, for $x \in X$ *fixed*, we deduce that

$$\lim_{n\to\infty} \langle P(E_n)x, x' \rangle = 0, \qquad x' \in X'.$$

This shows that the X-valued function $m := P(\cdot)x : \Sigma \longrightarrow X$ has the property that $\langle m, x' \rangle$ is σ-additive for all $x' \in X'$. By Proposition I.1 we conclude that m is a vector measure and so, in particular, $P(E_n)x = m(E_n) \longrightarrow 0$ with respect to the norm topology in X. Since $x \in X$ is arbitrary, it follows that $P(E_n) \longrightarrow 0$ in $L_s(X)$. Hence, P is σ-additive in $L_s(X)$.

The uniqueness of P follows from the uniqueness statement in Theorem II.4 together with the fact that $Ba(\Omega_B) = \sigma(Co(\Omega_B))$.

(b) Modify the proof of part (a) using Theorem II.5 in place of Theorem II.4.   ∎

Lemma IV.3.   *Let X be a Banach space and $P : \Sigma \longrightarrow \mathcal{L}_s(X)$ be a spectral measure. Then the following statements are equivalent.*

(a) *P is a closed spectral measure (c.f. Definition IV.2).*

(b) *The B.a. $\Sigma(P)$ is abstractly complete and, whenever $\{[E_\alpha]\} \subseteq \Sigma(P)$ is a net downwards filtering to $[\emptyset]$, it follows that $\lim_\alpha P(E_\alpha) = 0$ in $\mathcal{L}_s(X)$.*

(c) *The B.a. $\Sigma(P)$ is abstractly complete and, whenever $\{[E_\alpha]\} \subseteq \Sigma(P)$ is a net downwards filtering to $[\emptyset]$, it follows that $\lim_\alpha P(E_\alpha) = 0$ in $\mathcal{L}_w(X)$.*

Proof.   We first show that $\mathcal{L}_s(X)$ is *quasicomplete*, meaning that every bounded Cauchy net $\{T_\alpha\} \subseteq \mathcal{L}_s(X)$ (i.e. for which $\sup_\alpha \|T_\alpha x\| < \infty$ for each $x \in X$) has a limit in $\mathcal{L}_s(X)$. But, for fixed $x \in X$, the definition of the topology in $\mathcal{L}_s(X)$ implies that the net $\{T_\alpha x\}$ is Cauchy in X and so has a limit, say Tx, in X (as X is complete). Then $T : X \longrightarrow X$ defined by $x \mapsto Tx$ is an element of $\mathcal{L}(X)$ and $T_\alpha \xrightarrow{\alpha} T$ in $\mathcal{L}_s(X)$; see Proposition III.1(b).

The equivalence (a)$\Longleftrightarrow$(b) now follows (by choosing $m := P$ and $Y := \mathcal{L}_s(X)$) from a general result about vector measures $m : \Sigma \longrightarrow Y$, where Y is a *quasicomplete* locally convex Hausdorff space, which states:

A vector measure $m : \Sigma \longrightarrow Y$ is closed if and only if $\Sigma(m)$ (identified with $\{\chi_E : E \in \Sigma\} \subseteq L^1(m)$) is complete as an abstract B.a. and whenever $\{[E_\alpha]_m\} \subseteq \Sigma(m)$ is downwards filtering to $[\emptyset]$, it follows that $m(E_\alpha) \xrightarrow{\alpha} 0$ in Y.

For the proof of this result and a more precise definition of the notation and concepts involved, we refer to [11; Proposition 1.1].

The equivalence (b)$\Longleftrightarrow$(c) follows from the fact that the locally convex Hausdorff space $\mathcal{L}_s(X)$ equipped with its weak topology $\sigma(\mathcal{L}_s(X), (\mathcal{L}_s(X))'$ is precisely the space $\mathcal{L}_w(X)$ (this is essentially Proposition III.1(c)), and a general fact about vector measures, [34; Proposition 2], which states:

Let Y be a locally convex Hausdorff space and $m : \Sigma \longrightarrow Y$ be a vector measure. Then m is a closed measure if and only if it is a closed measure for every locally convex topology on Y consistent with the duality between Y and Y'.   ∎

We can now formulate one of the main representation theorems for B.a.'s of projections (of a certain kind) in terms of spectral measures.

Theorem IV.2.   *Let X be a Banach space and $\mathcal{B} \subseteq \mathcal{L}(X)$ be a B.a. of projections.*

(a) *If $\mathcal{B}$ has the σ-monotone property, then there exists a unique spectral measure $P : Ba(\Omega_B) \longrightarrow \mathcal{L}(X)$, where Ω_B is the Stone space of $\mathcal{B}$, such that $P(Ba(\Omega_B)) = \mathcal{B}$ and P restricted to $Co(\Omega_B)$ coincides with the Stone map $Q : Co(\Omega_B) \longrightarrow \mathcal{B}$.*

(b) *If $\mathcal{B}$ has the monotone property, then there exists a unique closed spectral measure $P : Bo(\Omega_B) \longrightarrow \mathcal{L}(X)$, where Ω_B is the Stone space of $\mathcal{B}$, with the properties that $P(Bo(\Omega_B)) = \mathcal{B}$, the map P restricted to $Co(\Omega_B)$ coincides with the Stone map $Q : Co(\Omega_B) \longrightarrow \mathcal{B}$, and* the spectral measure P is regular in the sense that

$$P(E) = \lim_{V \in \mathcal{V}(E)} P(V), \qquad E \in Bo(\Omega_{\mathcal{B}}), \tag{1}$$

where the limit of the net in (1) exists in $\mathcal{L}_s(X)$ and $\mathcal{V}(E)$ is the downwards filtering system (directed by inclusion) of all open sets $V \subseteq \Omega$ such that $E \subseteq V$.

Proof. (a) This is Lemma IV.2(a).

(b) Lemma IV.1(a) implies that $\mathcal{B}$ is abstractly complete and so, by Lemma IV.2(b), there is a spectral measure $P : Bo(\Omega_{\mathcal{B}}) \longrightarrow \mathcal{L}_s(X)$ which extends the Stone map Q from $Co(\Omega_{\mathcal{B}})$ onto $Bo(\Omega_{\mathcal{B}})$ and satisfies $P(Bo(\Omega_{\mathcal{B}})) = \mathcal{B}$. For simplicity of notation let $\Sigma := Bo(\Omega_{\mathcal{B}})$. The quotient B.a. $\Sigma(P)$ is abstractly complete because it is isomorphic to the abstractly complete B.a. $\mathcal{B}$.

To deduce the closedness of P we wish to use Lemma IV.3. So, let $\{[E_\alpha]\} \subseteq \Sigma(P)$ be decreasing to $[\emptyset]$. Since the induced map $\widetilde{P} : \Sigma(P) \longrightarrow \mathcal{B}$ given by $\widetilde{P}([E]) = P(E)$ is a B.a. isomorphism it follows that $\{P(E_\alpha)\}$ is decreasing in $\mathcal{B}$ to $\wedge_\alpha P(E_\alpha) = \widetilde{P}(\wedge_\alpha[E_\alpha]) = \widetilde{P}([\emptyset]) = 0$. By Lemma IV.1(a) it follows that $P(E_\alpha) \xrightarrow{\alpha} 0$ in $\mathcal{L}_w(X)$. Hence, Lemma IV.3 does indeed apply to show that P is a *closed* spectral measure.

Let $\{P(F_\beta)\} \subseteq \mathcal{B}$ be an increasing net. Since $\widetilde{P} : \Sigma(P) \longrightarrow \mathcal{B}$ is a B.a. isomorphism the net $\{[F_\beta]\}$ is increasing in $\Sigma(P)$. Since $\Sigma(P)$ is abstractly complete there is $[F] \in \Sigma(P)$ such that $\vee_\beta[F_\beta] = [F]$ and hence, $\{[F\backslash F_\beta]\}$ is downwards filtering to $[\emptyset]$ in $\Sigma(P)$. By Lemma IV.3 we conclude that $P(F\backslash F_\beta) \xrightarrow{\beta} 0$ in $\mathcal{L}_s(X)$. Now, for each β,

$$P(F) - P(F_\beta) = (P(F\backslash F_\beta) + P(F \cap F_\beta)) - P(F_\beta) = P(F\backslash F_\beta) + (P(F \cap F_\beta) - P(F_\beta)). \tag{2}$$

But $P(F \cap F_\beta) - P(F_\beta) = 0$, since $[F_\beta] \uparrow [F]$ implies that $[F] \wedge [F_\beta] = [F_\beta]$, and so

$$P(F \cap F_\beta) = \widetilde{P}([F \cap F_\beta]) = \widetilde{P}([F] \wedge [F_\beta]) = \widetilde{P}([F_\beta]) = P(F_\beta).$$

It follows from (2) that $P(F_\beta) \xrightarrow{\beta} P(F)$ in $\mathcal{L}_s(X)$. Similarly, if $\{P(H_\gamma)\} \subseteq \mathcal{B}$ is a decreasing net, then $\{[H_\gamma]\}$ is decreasing in $\Sigma(P)$ and so by abstract completeness of $\Sigma(P)$ again there is $[H] \in \Sigma(P)$ such that $\{[H_\gamma\backslash H]\} \downarrow [\emptyset]$ in $\Sigma(P)$. Then Lemma IV.3 yields $P(H_\gamma) \xrightarrow{\gamma} P(H)$ in $\mathcal{L}_s(X)$. So, we have established that $\mathcal{B}$ has the ordered convergence property.

Suppose for the moment that P is regular. Then the uniqueness of P follows immediately. Indeed, let $R : Bo(\Omega_{\mathcal{B}}) \longrightarrow \mathcal{L}_s(X)$ be another spectral measure which coincides with Q on $Co(\Omega_{\mathcal{B}})$, has range $\mathcal{B}$ and is regular. By Lemma IV.1(a) $\mathcal{B}$ is abstractly complete and so R and P coincide on the open sets in $\Omega_{\mathcal{B}}$; see Theorem II.5. Then (1) implies that $R = P$ on $Bo(\Omega_{\mathcal{B}})$.

So, it remains to establish the regularity of P. Fix $x \in X$ and $x' \in X'$, and let μ be the real part of the complex measure $\langle Px, x' \rangle$, that is,

$$\mu(E) = \mathrm{Re}(\langle P(E)x, x' \rangle), \qquad E \in Bo(\Omega_{\mathcal{B}}).$$

Then μ is a σ-additive, $\mathbb{R}$-valued measure and so by the *Hahn decomposition theorem* [14; p.129] there are pairwise disjoint Borel sets E_1 and E_2 with $\Omega_B = E_1 \cup E_2$ and $\mu(E_j \cap E) \geq 0$, for all $E \in Bo(\Omega_B)$ and $j \in \{1, 2\}$, such that

$$\mu(E) = \mu(E \cap E_1) - \mu(E \cap E_2), \qquad E \in Bo(\Omega_B).$$

Let $\Psi : Bo(\Omega_B) \longrightarrow Co(\Omega_B)$ be the σ-homomorphism which maps each $E \in Bo(\Omega_B)$ to the unique set $\Psi(E) \in Co(\Omega_B)$ such that $E \triangle \Psi(E)$ is of the first category . Now Ψ is a B.a. homomorphism and so the sets $U_j := \Psi(E_j)$, $j = 1, 2$, are pairwise disjoint, belong to $Co(\Omega_B)$ and satisfy $U_1 \cap U_2 = \Omega_B$. Since $E_1 \triangle U_1$ and $E_2 \triangle U_2$ are both P-null we have

$$\mu(E) = \mu(E \cap U_1) - \mu(E \cap U_2), \qquad E \in Bo(\Omega_B).$$

Define non-negative measures μ_1 and μ_2 by

$$\mu_j(E) = \mu(E \cap U_j), \qquad E \in Bo(\Omega_B), \quad j \in \{1, 2\}.$$

Fix an open set $V \subseteq \Omega_B$ and let $\{V_\alpha\}_{\alpha \in A}$ be the family of all open sets in Ω_B such that $V_\alpha \subseteq V$, for $\alpha \in A$. Ordering $\{V_\alpha \cap U_j\}_{\alpha \in A}$ by inclusion we deduce that $\{P(V_\alpha \cap U_j)\}_{\alpha \in A}$ is increasing in $\mathcal{B}$, as P is a homomorphism , and that $P(V \cap U_j) = \vee_{\alpha \in A} P(V_\alpha \cap U_j)$ by Theorem II.5. Since $\mathcal{B}$ has the monotone property, it follows that $P(V_\alpha \cap U_j) \xrightarrow{\alpha} P(V \cap U_j)$ in $\mathcal{L}_w(X)$; see Lemma IV.1(a). Accordingly, $\langle P(V_\alpha \cup U_j)x, x' \rangle \xrightarrow{\alpha} \langle P(V \cap U_j)x, x' \rangle$. Since $z \mapsto Re(z)$ is a continuous function on $\mathbb{C}$ it follows that also $\mu(V_\alpha \cap U_j) \xrightarrow{\alpha} \mu(V \cap U_j)$, i.e. $\mu_j(V_\alpha) \xrightarrow{\alpha} \mu_j(V)$ for $j = 1, 2$. Accordingly,

$$\mu_j(V) = \sup_\alpha \mu_j(V_\alpha), \qquad V \subseteq \Omega_B, \ V \text{ open}, \ j \in \{1, 2\}.$$

That is, both μ_1 and μ_2 are τ-*additive measures*; see [20] for this notion. It then follows from [20; Theorem 5.4] that both μ_1 and μ_2 are *regular* Borel measures and hence, μ is also regular. By a similar argument the imaginary part of $\langle Px, x' \rangle$ is also a regular measure and hence, $\langle Px, x' \rangle$ itself is regular. Accordingly,

$$\langle Px, x' \rangle(E) = \lim_{V \in \mathcal{V}(E)} \langle Px, x' \rangle(V), \qquad E \in Bo(\Omega_B).$$

Since $x \in X$ and $x' \in X'$ are arbitrary we deduce that

$$(3) \qquad \qquad \lim_{V \in \mathcal{V}(E)} P(V) = P(E) \qquad (\text{in } \mathcal{L}_w(X))$$

for each $E \in Bo(\Omega_B)$. But, as already noted above, $\mathcal{B}$ has the ordered convergence property and so $T_E := \lim_{V \in \mathcal{V}(E)} P(V)$ exists in $\mathcal{L}_s(X)$. Then (3) implies that $T_E = P(E)$ and so (1) holds for every $E \in Bo(\Omega_B)$, that is, P is regular. ■

The notion of *regularity* used in the paper [20], which is cited in the proof of the previous theorem, appears different to that given in Definition I.2. The following exercise shows that the two notions of regularity are actually the *same*.

Exercise 40.   (a) Let $\Sigma := Bo(\Omega)$, where Ω is a compact Hausdorff space and $\nu : \Sigma \longrightarrow \mathbb{C}$ be a complex measure . Let $|\nu| : \Sigma \longrightarrow [0, \infty)$ be the variation measure of ν. Show, in the sense of Definition I.2, that ν is regular if and only if $|\nu|$ is regular.

(b) Let $\mu : Bo(\Omega) \longrightarrow [0, \infty)$ be a finite, σ-additive measure. For each subset $E \subseteq \Omega$ define

$$\mu_i(E) := \sup\{\mu(Z) : Z \subseteq E,\ Z \text{ closed in } \Omega\}$$

and
$$\mu_e(E) := \inf\{\mu(U) : E \subseteq U,\ U \text{ open in } \Omega\}.$$
Show that the following statements are equivalent.

(i) μ is regular.

(ii) $\mu(B) = \mu_i(B)$, for every $B \in Bo(\Omega)$.

(iii) $\mu(B) = \mu_e(B)$, for every $B \in Bo(\Omega)$.

Deduce that regularity as given in [20] coincides with that of Definition I.2. ∎

We are now ready to establish the main result of this chapter.

Proof of Theorem IV.1. (b)$\Rightarrow$(a). This is clear as convergence of nets (and sequences) in $\mathcal{L}_s(X)$ also implies their convergence in $\mathcal{L}_w(X)$ and to the same limit.

(a)$\Rightarrow$(d). This is immediate from Theorem IV.2(a) if $\mathcal{B}$ has the σ-monotone property and from Theorem IV.2(b) if $\mathcal{B}$ has the monotone property.

(c)$\Rightarrow$(b). Suppose that $\mathcal{B}$ is Bade complete . Let $\{B_\alpha\} \subseteq \mathcal{B}$ be an increasing net and let $B = \vee_\alpha B_\alpha$. Fix $x \in X$. Given $\varepsilon > 0$, the fact that $Bx \in BX = \overline{\text{sp}}\{\cup_\alpha B_\alpha X\}$ means there is a vector $y = \sum_{j=1}^n z_j$ and indices α_j such that $z_j \in B_{\alpha_j}X$ (i.e. $z_j = B_{\alpha_j}z_j$), for $1 \leq j \leq n$, and $\|y - Bx\| \leq \varepsilon$. For each $\alpha \geq \alpha_j$ $(1 \leq j \leq n)$ we have $B_\alpha y = y$ and $B_\alpha B = B_\alpha$ (as $B_\alpha \uparrow B$), and hence

$$\|B_\alpha x - Bx\| \leq \|B_\alpha x - y\| + \|y - Bx\| = \|B_\alpha(Bx - y)\| + \|y - Bx\|$$
$$\leq (\|B_\alpha\| + 1)\|y - Bx\| \leq (1 + \|\mathcal{B}\|)\varepsilon,$$

where $\|\mathcal{B}\| < \infty$ (see Theorem III.1). This shows that $\lim_\alpha B_\alpha x = Bx$, for the norm topology in X. Since $x \in X$ is arbitrary, we have shown that $\lim_\alpha B_\alpha = B$ in $\mathcal{L}_s(X)$.

The dual statement for decreasing nets follows from the formula $\wedge_\alpha B_\alpha = I - \vee_\alpha(I - B_\alpha)$. So, $\mathcal{B}$ has the ordered convergence property.

The proof when $\mathcal{B}$ is Bade σ-complete is similar; just replace nets by sequences.

(b)$\Rightarrow$(c).   Suppose that $\mathcal{B}$ has the ordered convergence property. By Lemma IV.1(b) $\mathcal{B}$ is abstractly complete as a B.a. Let $\mathcal{A} \subseteq \mathcal{B}$ be a set and let $\{B_\alpha\}$ be the increasing net consisting of the suprema of all finite subsets of $\mathcal{A}$, directed by the order induced from $\mathcal{B}$. Then an element of $\mathcal{B}$ is an upper bound for $\mathcal{A}$ if and only if it is an upper bound for $\{B_\alpha\}$. Since $(F_1 \vee \ldots \vee F_n)X = \overline{\text{sp}}\{\cup_{j=1}^n F_j X\}$ for any finite set $\{F_j\}_{j=1}^n \subseteq \mathcal{A}$, to construct a least upper bound for $\mathcal{A}$ with the property required in the definition of Bade completeness (c.f. Definition IV.1(i)) it suffices to make the corresponding construction for $\{B_\alpha\}$. Now by Lemma IV.1(b) we have $\vee_\alpha B_\alpha = B$, where $B = \lim_\alpha B_\alpha$ in $\mathcal{L}_s(X)$. So, it remains to check that

(4) $$BX = \overline{\text{sp}}\{\cup_\alpha B_\alpha X\}.$$

Fix $x \in X$. Since $B_\alpha x \xrightarrow{\alpha} Bx$ in X and $B_\alpha x \in \overline{\text{sp}\{\cup_\gamma B_\gamma X\}}$, for all α, it follows that $Bx \in \overline{\text{sp}\{\cup_\alpha B_\alpha X\}}$. This shows that $BX \subseteq \overline{\text{sp}\{\cup_\alpha B_\alpha X\}}$. On the other hand, since $\{B_\alpha\}$ is increasing with $\vee_\alpha B_\alpha = B$ we have $B_\alpha B = B_\alpha$ i.e. $B_\alpha X \subseteq BX$, for all α, and so $\overline{\text{sp}\{\cup_\alpha B_\alpha X\}} \subseteq BX$ follows. So, (4) is indeed satisfied.

By considering the decreasing net $\{I - B_\alpha\}$, a greatest lower bound for $\mathcal{A}$ with the property required in Definition IV.1(i) can be constructed in a similar way as the least upper bound was constructed. Hence, $\mathcal{B}$ is Bade complete .

The proof when $\mathcal{B}$ has the σ-ordered convergence property is similar; just replace nets by sequences.

(d)$\Rightarrow$(b). If $\mathcal{B}$ coincides with the range of a spectral measure, then $\mathcal{B}$ has the σ-ordered convergence property; this was established in the proof of Lemma IV.2(a).

If $\mathcal{B}$ coincides with the range of a *closed* spectral measure , then $\mathcal{B}$ has the ordered convergence property; this was established in the proof of Theorem IV.2(b). ∎

We conclude with an exercise showing that a finitely additive spectral measure can be σ-additive in $\mathcal{L}_u(X)$ only in trivial cases.

Exercise 41.   Let X be an infinite dimensional Banach space.

(a) Let $R \in \mathcal{L}(X)$ be a non-zero projection. Show that $\|R\| \geq 1$.

(b) Let (Ω, Σ) be a measurable space and $P : \Sigma \longrightarrow \mathcal{L}(X)$ be a multiplicative set function which is σ-additive in the Banach space $\mathcal{L}_u(X)$, that is, $\lim_{n\to\infty} \|P(E_n)\| = 0$ whenever $\{E_n\}_{n=1}^\infty \subseteq \Sigma$ is a sequence decreasing to $\emptyset$; see Exercise 9(b). Show that there exists $N \in \mathbb{N}$ such that $P(E_n) = 0$ for all $n \geq N$.

(c) Let $P : \Sigma \longrightarrow \mathcal{L}(X)$ be as in part (b). Show that the range $P(\Sigma)$, of P, is a finite subset of $\mathcal{L}(X)$. ∎

Chapter V

Integral representation of the strongly closed algebra generated by a Boolean algebra of projections

Let $\mathcal{B}$ be a Bade complete B.a. of projections in a Banach space X. We have seen that there always exists some closed spectral measure $P : \Sigma \longrightarrow \mathcal{L}_s(X)$ such that $P(\Sigma) = \mathcal{B}$; for instance, Σ can always be taken to be $Bo(\Omega_{\mathcal{B}})$, where $\Omega_{\mathcal{B}}$ is the Stone space of $\mathcal{B}$. Associated with P is the space $L^1(P)$ of all (equivalence classes of) P-integrable functions. It will be shown that $L^1(P)$ can be topologized (via a *non-normable* topology) in such a way that $L^1(P)$ is a *complete* locally convex Hausdorff space and the (linear) integration map $f \mapsto \int_\Omega f \, dP$ is a bicontinuous topological and algebraic isomorphism of $L^1(P)$ onto the closed subalgebra $\langle \mathcal{B} \rangle_s^-$, of $\mathcal{L}_s(X)$, generated by $\mathcal{B}$. The *closedness* of P turns out to play an essential role in this respect. In particular, $L^1(P)$ itself turns out to be a commutative *algebra* of functions!

In the first part of this chapter we develop the theory of integration with respect to spectral measures and investigate the space $L^1(P)$, but only as far as is needed to establish the integral representation theorem mentioned above. A fundamental result is the fact that the only P-integrable functions are the P-essentially bounded functions! Having established the representation theorem identifying $\langle \mathcal{B} \rangle_s^-$ as an $L^1(P)$-space, the remainder of the chapter concentrates on highlighting various non-trivial consequences of this theorem concerning Bade complete B.a. 's of projections $\mathcal{B}$ and the subalgebras $\langle \mathcal{B} \rangle_s^-$ that they generate. Many of these results can be found in [15; Chapter XVII, Section 3]. However, the proofs given there are, in many cases, quite different to those given here. This is not surprising since the approach adopted here, via the integral representation of the algebra $\langle \mathcal{B} \rangle_s^-$ as an $L^1(P)$-space, was not available in the late 1960's (when [15] was written); it appeared for the first time, some 15 years later, in [11] and was further exploited in [12]. The explanation for this delay lies in the observation that the theory of vector measures and integration was only systematically developed throughout the 1970's and 80's; see [8] and [27].

We begin by studying, in detail, the space of all functions which are integrable with respect to a given spectral measure $P : \Sigma \longrightarrow \mathcal{L}_s(X)$, where X is a Banach space and Σ is a

σ-algebra of subsets of some non-empty set Ω. According to Definition III.4 a Σ-measurable function $f : \Omega \longrightarrow \mathbb{C}$ is P-integrable if

(i) $f \in L^1(\langle Px, x' \rangle)$, for each $x \in X$ and $x' \in X'$,

and

(ii)   for each $E \in \Sigma$ there exists an operator $\int_E f dP \in \mathcal{L}(X)$ satisfying

$$\langle (\int_E f \, dP)x, x' \rangle = \int_E f \, d\langle Px, x' \rangle, \qquad x \in X, \, x' \in X'.$$

If $\mathcal{L}_s(X)$ is considered as a *locally convex Hausdorff space*, where the topology is specified by the family of *seminorms* $\{q_x : x \in X\}$ given by

$$q_x(T) := \|Tx\|, \qquad T \in \mathcal{L}(X),$$

for each $x \in X$, then the above definition of P-integrability is a particular case of the definition of integration with respect to more general vector measures as developed in the book [27]. Because of the *multiplicativity* of P it turns out that the definition of P-integrability can be somewhat simplified; see Proposition V.1 below. We begin with a technical result.

Lemma V.1.   *Let X be a Banach space and $P : \Sigma \longrightarrow \mathcal{L}_s(X)$ be a spectral measure . Let $f : \Omega \longrightarrow \mathbb{C}$ be a Σ-measurable function such that $f \in L^1(\langle Px, x' \rangle)$, for all $x \in X$ and $x' \in X'$. If $T \in \mathcal{L}(X)$ satisfies*

$$\langle Tx, x' \rangle = \int_\Omega f \, d\langle Px, x' \rangle, \qquad x \in X, \, x' \in X',$$

then $TP(E) = P(E)T$, for all $E \in \Sigma$.

Proof.   Fix $E \in \Sigma$ and $x \in X$, $x' \in X'$. Then

$$\langle TP(E)x, x' \rangle = \langle Ty, x' \rangle = \int_\Omega f \, d\langle Py, x' \rangle,$$

where $y := P(E)x$. Also, if $\xi := P(E)'x'$ with $P(E)' \in \mathcal{L}(X')$ the dual operator , then

$$\langle P(E)Tx, x' \rangle = \langle Tx, \xi \rangle = \int_\Omega f \, d\langle Px, \xi \rangle.$$

So, it suffices to check that $\langle Py, x' \rangle$ and $\langle Px, \xi \rangle$ are equal as measures on Σ. But, for any $F \in \Sigma$,

$$\langle Py, x' \rangle(F) := \langle P(F)y, x' \rangle = \langle P(F)P(E)x, x' \rangle$$

and
$$\langle Px, \xi \rangle(F) := \langle P(F)x, \xi \rangle = \langle P(F)x, P(E)'x' \rangle = \langle P(E)P(F)x, x' \rangle. \qquad \blacksquare$$

Proposition V.1.   *Let X be a Banach space and $P : \Sigma \longrightarrow \mathcal{L}_s(X)$ be a spectral measure . Let $f : \Omega \longrightarrow \mathbb{C}$ be a Σ-measurable function such that $f \in L^1(\langle Px, x' \rangle)$ for every $x \in X$ and $x' \in X'$. Then f is P-integrable if and only if there exists $T \in \mathcal{L}(X)$ satisfying*

(1) $$\langle Tx, x' \rangle = \int_\Omega f \, d\langle Px, x' \rangle, \qquad x \in X, \, x' \in X'.$$

In this case

$$(2) \qquad \int_E f \, dP = TP(E) = P(E)T, \qquad E \in \Sigma,$$

and, in particular, $T = \int_\Omega f \, dP$.

Proof. Suppose that an operator $T \in \mathcal{L}(X)$ exists satisfying (1). *Define $\int_E f \, dP :=$ $TP(E) (= P(E)T$ by Lemma V.1), for each $E \in \Sigma$. Then, given $x \in X$ and $x' \in X'$,* we have for fixed $E \in \Sigma$ that

$$\left\langle \left(\int_E f \, dP \right)x, x' \right\rangle = \langle TP(E)x, x' \rangle = \int_\Omega f \, d\langle Py, x' \rangle$$

where $y := P(E)x$. But, $\langle Py, x' \rangle$ is the complex measure on Σ given by

$$\langle Py, x' \rangle(F) := \langle P(F)y, x' \rangle = \langle P(F)P(E)x, x' \rangle = \langle P(E \cap F)x, x' \rangle$$
$$\int_\Omega \chi_{E \cap F} \, d\langle Px, x' \rangle = \int_\Omega \chi_E \chi_F \, d\langle Px, x' \rangle = \int_F \chi_E \, d\langle Px, x' \rangle,$$

for $F \in \Sigma$. That is, $\langle Py, x' \rangle(F) = \int_F \chi_E \, d\langle Px, x' \rangle$, for $F \in \Sigma$, and so

$$\int_\Omega f \, d\langle Py, x' \rangle = \int_\Omega f \chi_E \, d\langle Px, x' \rangle = \int_E f \, d\langle Px, x' \rangle.$$

That is, for each $E \in \Sigma$, we have

$$\left\langle \left(\int_E f \, dP \right)x, x' \right\rangle = \int_E f \, d\langle Px, x' \rangle, \qquad x \in X, \, x' \in X'.$$

This means by definition that f is P-integrable .

Conversely, if f is P-integrable, then the operator $T := \int_\Omega f \, dP \in \mathcal{L}(X)$ satisfies (1) by definition of P-integrability. ∎

Let X be a Banach space and $P : \Sigma \longrightarrow \mathcal{L}_s(X)$ be a spectral measure . Then the vector space of all P-integrable functions is denoted by $L(P)$. Given $f \in L(P)$, the operator $\int_\Omega f \, dP \in \mathcal{L}(X)$ is also denoted by $P(f)$; it necessarily satisfies (1) with $T = P(f)$ and commutes with each projection $P(E)$, for $E \in \Sigma$.

The next result clarifies the relationship between P-integrable functions and those functions which are integrable with respect to each X-valued vector measure Px, for $x \in X$. This result is very useful in practice since it reduces the problem of determining whether a function is integrable with respect to an $\mathcal{L}_s(X)$-valued measure to verifying whether it is integrable with respect to a family of X-valued measures.

Proposition V.2. *Let X be a Banach space and $P : \Sigma \longrightarrow \mathcal{L}_s(X)$ be a spectral measure . Let $f : \Omega \longrightarrow \mathbb{C}$ be a Σ-measurable function. Then f is P-integrable if and only if f is Px-integrable for each $x \in X$, that is,*

$$L(P) = \bigcap_{x \in X} L(Px).$$

Proof. Suppose that f is P-integrable. Fix $x \in X$. Since $f \in L^1(\langle Pu, u' \rangle)$ for all $u \in X$ and $u' \in X'$ it is clear that $f \in L^1(\langle m, x' \rangle)$ for all $x' \in X'$, where $m := Px$. For each $E \in \Sigma$ define $\int_E f \, dm := (\int_E f dP)x$. Then, using the fact that f is P-integrable, gives

$$\langle \int_E f \, dm, x' \rangle = \langle (\int_E f \, dP)x, x' \rangle = \int_E f \, d\langle Px, x' \rangle = \int_E f \, d\langle m, x' \rangle,$$

for each $x' \in X'$. So, f is Px-integrable and satisfies

(3)
$$\int_E f \, d(Px) = (\int_E f \, dP)x, \qquad E \in \Sigma.$$

Conversely, suppose that $f \in L(Px)$, for each $x \in X$. Given $n \in \mathbb{N}$, let $E(n) := \{w \in \Omega : |f(w)| \le n\}$. Then $f_n := f \chi_{E(n)}$, for $n \in \mathbb{N}$, is a bounded Σ-measurable function and hence $f_n \in L(P)$; see Lemma III.3 and Proposition III.2. Fix $x \in X$. By the first part of the proof we know that $f_n \in L(Px)$ and

$$\int_\Omega f_n \, d(Px) = (\int_\Omega f_n \, dP)x, \qquad n \in \mathbb{N}.$$

Since $|f_n| \le |f|$ with $f \in L(Px)$ and $f_n \longrightarrow f$ pointwise on Ω we can apply the dominated convergence theorem for the X-valued vector measure Px (see Theorem I.9) to deduce that

$$Tx := \lim_{n \to \infty} (\int_\Omega f_n \, dP)x = \lim_{n \to \infty} \int_\Omega f_n \, d(Px) = \int_\Omega f \, d(Px)$$

exists in the norm of X. In particular, $\sup_n \|(\int_\Omega f_n \, dP)x\| < \infty$ and so $\sup_n \| \int_\Omega f_n \, dP \| < \infty$ by Proposition III.1(a). Since $x \in X$ is arbitrary and $\{\int_\Omega f_n \, dP\}_{n=1}^\infty \subseteq \mathcal{L}(X)$ it follows from the Banach-Steinhaus theorem (c.f. Proposition III.1(b)) that the so defined operator $T \in \mathcal{L}(X)$. It is routine to verify that T satisfies (1) of Proposition V.1 and so, by that result, f is P-integrable . ∎

Exercise 42. A Banach space X is said to have the *BP-property* if, whenever $\{x_n\}_{n=1}^\infty$ is a sequence in X satisfying $\sum_{n=1}^\infty |\langle x_n, x' \rangle| < \infty$ for each $x' \in X'$, then the series $\sum_{n=1}^\infty x_n$ is convergent in norm to some element of X. It is a classical theorem that X has the *BP*-property if and only if X does not have any closed subspace which is isomorphic (as a Banach space) to c_0; see [3].

(a) Let X be a Banach space with the *BP*-property and $m : \Sigma \longrightarrow X$ be any vector measure. Show that a Σ-measurable function f is m-integrable if and only if

(∗)
$$\int_\Omega |f| \, d|\langle m, x' \rangle| < \infty, \qquad x' \in X'.$$

(b) Give an example of a vector measure $m : \Sigma \longrightarrow c_0$, defined on some σ-algebra Σ, and a Σ-measurable function f satisfying (∗) of part (a) for every $x' \in (c_0)' = \ell^1$, such that f is *not* m-integrable.

(c) Let X be a Banach space with the BP-property and $P : \Sigma \longrightarrow \mathcal{L}_s(X)$ be a spectral measure . Show that a Σ-measurable function f is P-integrable if and only if

$$\int_\Omega |f| \, d|\langle Px, x' \rangle| < \infty, \qquad x \in X, \, x' \in X'. \qquad \blacksquare$$

The next two results show that Proposition V.2 has some useful consequences.

Corollary V.2.1. *Let X be a Banach space and $P : \Sigma \longrightarrow \mathcal{L}_s(X)$ be a spectral measure .*

(a) *If $f \in L(P)$, then also $|f| \in L(P)$.*

(b) *If $0 \le g \in L(P)$ and $f : \Omega \longrightarrow \mathbb{C}$ is a measurable function such that $|f| \le g$, then $f \in L(P)$.*

Proof. (a) By Proposition V.2 we have that $f \in L(Px)$ for all $x \in X$ and hence, that $|f| \in L(Px)$ for all $x \in X$ (see Lemma I.3). Then $|f|$ is P-integrable by Proposition V.2 (again!).

(b) Since $g \in L(Px)$ for all $x \in X$ (by Proposition V.2), it follows that $f \in L(Px)$ for all $x \in X$ (by Lemma I.4) and so Proposition V.2 implies that $f \in L(P)$. $\qquad \blacksquare$

Corollary V.2.2. *(Dominated convergence theorem for spectral measures). Let X be a Banach space and $P : \Sigma \longrightarrow \mathcal{L}_s(X)$ be a spectral measure. Let $f_n : \Omega \longrightarrow \mathbb{C}$, for $n \in \mathbb{N}$, be a sequence of Σ-measurable functions such that $f(w) := \lim_{n \to \infty} f_n(w)$ exists pointwise on Ω and $|f_n| \le g$, for all $n \in \mathbb{N}$ and some $0 \le g \in L(P)$. Then $f \in L(P)$ and $P(f_n) \longrightarrow P(f)$ in $\mathcal{L}_s(X)$.*

Proof. Fix $x \in X$. Then $g \in L(Px)$; see Proposition V.2. It follows from the dominated convergence theorem for the X-valued measure Px (see Theorem I.9) that $f \in L(Px)$ and $\int_\Omega f_n \, d(Px) \longrightarrow \int_\Omega f \, d(Px)$ in X. That is, $(\int_\Omega f_n \, dP)x \longrightarrow \int_\Omega f \, d(Px)$, where we have used Corollary V.2.1 to ensure that $f_n \in L(P)$ for each $n \in \mathbb{N}$. Then Proposition V.2 again gives that $f \in L(P)$ and so

$$\left(\int_\Omega f_n \, dP \right)x \longrightarrow \int_\Omega f \, d(Px) = \left(\int_\Omega f \, dP \right)x, \qquad n \to \infty,$$

in the norm of X. Since $x \in X$ is arbitrary, we see that $P(f_n) \longrightarrow P(f)$ in $\mathcal{L}_s(X)$. $\qquad \blacksquare$

The next result shows that $L(P)$ is an *algebra* with respect to pointwise multiplication of functions. It should be noted that this feature is special to *spectral measures* and is surely not typical of more general operator-valued measures. The *multiplicativity* of P plays an essential role in this regard.

Proposition V.3. *Let X be a Banach space and $P : \Sigma \longrightarrow \mathcal{L}_s(X)$ be a spectral measure. If f and g are P-integrable functions, then fg is also P-integrable . Moreover,*

$$(4) \qquad \int_E (fg) dP = P(f)P(g)P(E) = P(g)P(f)P(E), \qquad E \in \Sigma.$$

Proof. It is a simple consequence of (1) and (2) that (4) is valid whenever $g \in \text{sim}(\Sigma)$ and $f \in L(P)$.

Suppose that $f \in L(P)$ and g is bounded and Σ-measurable. Then $|fg| \le \|g\|_\infty \cdot |f|$ and so Corollary V.2.1 yields that $fg \in L(P)$. Select functions $\{s_n\}_{n=1}^\infty \subseteq \text{sim}(\Sigma)$ with

$|s_n| \leq |g|$ such that $s_n \longrightarrow g$ uniformly on Ω. Since $P(f)P(s_n) = P(fs_n)$, for $n \in \mathbb{N}$, with $|s_n| \leq |g| \in L(P)$ and $|fs_n| \leq |f| \cdot \|g\|_\infty \in L(P)$ with $fs_n \longrightarrow fg$ pointwise on Ω, it follows from Corollary V.2.2 that both $P(s_n) \longrightarrow P(g)$ and $P(fs_n) \longrightarrow P(fg)$ in $\mathcal{L}_s(X)$. In particular, $P(f)P(g) = P(fg)$ follows, that is, again (4) holds.

Suppose now that both $f, g \in L(P)$ are arbitrary. Let $E(n) := \{w : |g(w)| \leq n\}$, for $n \in \mathbb{N}$, and define $g_n := g\chi_{E(n)}$. Then $fg_n \longrightarrow fg$ pointwise on Ω and

$$(5) \qquad \int_E fg_n \, d\langle Px, x'\rangle = \int_\Omega fg\chi_{E \cap E(n)} \, d\langle Px, x'\rangle \qquad E \in \Sigma,$$

for all $n \in \mathbb{N}$ and $x \in X$, $x' \in X'$. Fix $E \in \Sigma$. Since $g\chi_{E \cap E(n)}$ is bounded for each $n \in \mathbb{N}$, it follows that $|fg\chi_{E \cap E_n}| \leq \|g\chi_{E \cap E(n)}\|_\infty \cdot |f|$ and so, by Corollary V.2.1, $fg\chi_{E \cap E(n)} \in L(P)$ with

$$(6) \qquad P(fg\chi_{E \cap E(n)}) = P(f)P(g\chi_{E \cap E(n)}), \qquad n \in \mathbb{N}.$$

A routine calculation using (1) and (6) shows that

$$\int_\Omega fg\chi_{E \cap E(n)} \, d\langle Px, x'\rangle = \int_E g\chi_{E(n)} \, d\langle Px, \xi\rangle, \qquad n \in \mathbb{N},$$

for each $x \in X$, $x' \in X'$, where $\xi := P(f)'x'$. Since $g \in L(P)$ we know that $g \in L^1(\langle Px, \xi\rangle)$. Moreover, since $|g\chi_{E(n)}| \leq |g|$ with $g\chi_{E(n)} \longrightarrow g$ pointwise on Ω, the dominated convergence theorem for the complex measure $\langle Px, \xi\rangle$ yields

$$\lim_{n \to \infty} \int_E g\chi_{E(n)} \, d\langle Px, \xi\rangle = \int_E g \, d\langle Px, \xi\rangle.$$

The existence of this limit and the identities (5) show that $\{\int_E fg_n \, d\langle Px, x'\rangle\}_{n=1}^\infty$ is a Cauchy sequence in $\mathbb{C}$, for each $E \in \Sigma$. Hence, fg is $\langle Px, x'\rangle$-integrable for each $x' \in X'$, $x \in X$; see Proposition I.6.(b).

Let $T := P(f)P(g)$. Then it follows, for $x \in X$ and $x' \in X'$, that

$$(7) \qquad \langle Tx, x'\rangle = \langle P(g)x, P(f)'x'\rangle = \int_\Omega g \, d\langle Px, P(f)'x'\rangle.$$

Since $E \mapsto \langle P(E)x, P(f)'x'\rangle$, for $E \in \Sigma$, coincides with the measure $E \mapsto \int_E f \, d\langle Px, x'\rangle$, for $E \in \Sigma$, it follows from (7) that

$$\langle Tx, x'\rangle = \int_\Omega fg \, d\langle Px, x'\rangle, \qquad x \in X, \, x' \in X'.$$

Proposition V.1 then implies that $fg \in L(P)$ and (4) is valid. ∎

We now turn out attention to another *algebra* of P-integrable functions.

Definition V.1. Let X be a Banach space and $P : \Sigma \longrightarrow \mathcal{L}_s(X)$ be a spectral measure. A P-integrable function $f : \Omega \longrightarrow \mathbb{C}$ is called *P-null* if $P(f) := \int_\Omega f \, dP = 0$. By Proposition V.1 this is equivalent to

$$\int_E f \, dP = 0, \qquad E \in \Sigma.$$

∎

Let $P : \Sigma \longrightarrow \mathcal{L}_s(X)$ be a spectral measure. Two P-integrable functions f and g are called P-*equivalent* if $|f - g|$ is P-null. This is the same as $\{w : f(w) \neq g(w)\}$ being a P-null set; see the discussion prior to Exercise 37 (in Chapter IV).

Definition V.2. Let $P : \Sigma \longrightarrow \mathcal{L}_s(X)$ be a spectral measure. Then a Σ-measurable function $f : \Omega \longrightarrow \mathbb{C}$ is said to be P-*essentially bounded* on Ω if

$$|f|_P := \inf\{\sup\{|f(w)| : w \in E\} : E \in \Sigma, \ P(E) = I\} < \infty. \qquad \blacksquare$$

Exercise 43. Let X be a Banach space and $P : \Sigma \longrightarrow \mathcal{L}_s(X)$ be a spectral measure. Let $f : \Omega \longrightarrow \mathbb{C}$ be a P-essentially bounded function. Show that there exists a set $E_0 \in \Sigma$ such that $P(E_0) = I$ and

$$|f|_P = \sup\{|f(w)| : w \in E_0\}. \qquad \blacksquare$$

(8) Let $P : \Sigma \longrightarrow \mathcal{L}_s(X)$ be a spectral measure and f be a P-essentially bounded function. By Exercise 43 there is a set $E_0 \in \Sigma$ with $P(E_0) = I$ such that (8) holds. Hence, there is a bounded Σ-measurable function f_0 on Ω (e.g. $f_0 = f\chi_{E_0}$) such that $\{w : f_0(w) \neq f(w)\}$ is P-null and $|f|_P = \|f_0\|_\infty$. We define the equivalence class $[f]$ of f to consist of all Σ-measurable functions g on Ω for which $\{w \in \Omega : f(w) \neq g(w)\}$ is a P-null set. The space of all equivalence classes of P-essentially bounded functions is denoted by $L^\infty(P)$; it is a Banach algebra with respect to the P-essential supremum norm $|\cdot|_P$. If $[f] \in L^\infty(P)$, then the integrals $\int_E [f]\, dP$ are defined to be $\int_E f_0\, dP$, $E \in \Sigma$, for any bounded Σ-measurable function $f_0 : \Omega \longrightarrow \mathbb{C}$ such that $f = f_0$, P-a.e.. By the usual abuse of notation we will write f rather than $[f]$ for elements of $L^\infty(P)$. Note that Lemma III.3 and Proposition III.2 imply that f_0 is P-integrable .

Exercise 44. Let X be a Banach space and $P : \Sigma \longrightarrow \mathcal{L}_s(X)$ be a spectral measure .

(a) Show that $L^\infty(P)$ is *complete* with respect to the norm $|\cdot|_P$.

(b) Show that the Σ-simple functions are dense in $L^\infty(P)$. $\qquad \blacksquare$

Theorem V.1. *Let X be a Banach space and $P : \Sigma \longrightarrow \mathcal{L}_s(X)$ be a spectral measure . Define $\Psi : L^\infty(P) \longrightarrow \mathcal{L}(X)$ by*

$$\Psi(f) := \int_\Omega f\, dP, \qquad f \in L^\infty(P).$$

Then Ψ is an isomorphism of the Banach algebra $L^\infty(P)$ onto an inverse closed Banach subalgebra of $\mathcal{L}_u(X)$ and satisfies

$$|f|_P \leq \|\Psi(f)\| \leq 4\|P(\Sigma)\| \cdot |f|_P, \qquad f \in L^\infty(P).$$

Proof. It is clear that Ψ is linear and multiplicative (by Proposition V.3). Let $f \in L^\infty(P)$ and choose $f_0 : \Omega \longrightarrow \mathbb{C}$ bounded and Σ-measurable such that $f = f_0$, P-a.e., and $|f|_P = \|f_0\|_\infty = \sup\{|f_0(w)| : w \in \Omega\}$. Choose functions $s_n \in \text{sim}(\Sigma)$ such that $|s_n| \leq |f_0|$ and $s_n \longrightarrow f_0$ uniformly on Ω. By the proof of Proposition III.2 we have, for each $n \in \mathbb{N}$, that

$$(9) \qquad \left\| \int_\Omega s_n\, dP \right\| \leq 4\|P(\Sigma)\| \cdot \|s_n\|_\infty \leq 4\|P(\Sigma)\| \cdot \|f_0\|_\infty = 4\|P(\Sigma)\| \cdot |f|_P,$$

and $\int_\Omega s_n \, dP \longrightarrow \int_\Omega f_0 \, dP$ in $\mathcal{L}_u(X)$. Hence, also

$$\left\| \int_\Omega s_n \, dP \right\| \longrightarrow \left\| \int_\Omega f_0 \, dP \right\| = \|\Psi(f)\|, \qquad n \to \infty,$$

and it follows from (9) that $\|\Psi(f)\| \le 4\|P(\Sigma)\| \cdot |f|_P$.

Let $f = \sum_{j=1}^n \alpha_j \chi_{E_j}$ with $E_j \cap E_k = \emptyset$ if $j \ne k$ and $\cup_{j=1}^n E_j = \Omega$. Then $|f|_P = \max\{|\alpha_j| : P(E_j) \ne 0\} = |\alpha_{j_0}|$ say (and assume $|f|_P > 0$). Since $P(E_{j_0}) \ne 0$ there is a unit vector $x \in X$ with $P(E_{j_0})x = x$. Then

$$\Psi(f)x = \sum_{j=1}^n \alpha_j P(E_j)x = \sum_{j=1}^n \alpha_j P(E_j)P(E_{j_0})x = \alpha_{j_0}x$$

and so $\|\Psi(f)\| \ge \|\Psi(f)x\| = |\alpha_{j_0}| = |f|_P$. Hence,

$$(10) \qquad\qquad\qquad |f|_P \le \|\Psi(f)\|, \qquad f \in \operatorname{sim}(\Sigma).$$

By Exercise 44 the space $\operatorname{sim}(\Sigma)$ is dense in $L^\infty(P)$ and so a continuous extension argument shows that (10) holds for all $f \in L^\infty(P)$.

To complete the proof it remains to check that the range $\Psi(L^\infty(P))$ is inverse closed in $\mathcal{L}_u(X)$. So, let $f \in L^\infty(P)$ and suppose that $(\Psi(f))^{-1}$ exists in $\mathcal{L}(X)$. For each $m \in \mathbb{N}$ let $\delta_1, \ldots, \delta_{n_m}$ be disjoint Borel sets (in $\mathbb{C}$) of diameter less than $1/m$ whose union is the compact disc $\{z \in \mathbb{C} : |z| \le |f|_P\}$. Let $E_j := f^{-1}(\delta_j)$ and choose any point $w_j \in E_j$. Then the Σ-simple functions

$$f_m := \sum_{j=1}^{n_m} f(w_j)\chi_{E_j}, \qquad m \in \mathbb{N},$$

have the properties

$$|f - f_m|_P \le (1/m) \quad \text{and} \quad \|\Psi(f) - \Psi(f_m)\| \le 4\|P(\Sigma)\|/m, \qquad m \in \mathbb{N}.$$

Then $\Psi(f_m) \longrightarrow \Psi(f)$ in $\mathcal{L}_u(X)$. Since $\Psi(f)$ is invertible in $\mathcal{L}(X)$, it follows from Lemma III.1 that for sufficiently large values of m each operator

$$\Psi(f_m) = \sum_{j=1}^{n_m} f_m(w_j)P(E_j)$$

is invertible in $\mathcal{L}(X)$. So, for all sufficiently large m we must have that $f_m(w_j) \ne 0$ whenever $P(E_j) \ne 0$, which means precisely that $(1/f_m) \in L^\infty(P)$ and, of course, $\Psi(1/f_m) = (\Psi(f_m))^{-1}$. By Exercise 27

$$|(1/f_m) - (1/f_k)|_P \le \|\Psi(1/f_m) - \Psi(1/f_k)\| = \|(\Psi(f_m))^{-1} - (\Psi(f_k))^{-1}\| \longrightarrow 0,$$

as $k, m \to \infty$. Accordingly, $\{1/f_m\}_{m=1}^{\infty}$ is Cauchy in $L^{\infty}(P)$ and so converges (to $1/f$ of course). Hence, $(1/f) \in L^{\infty}(P)$ and

$$(\Psi(f))^{-1} = \lim_{m \to \infty} (\Psi(f_m))^{-1} = \lim_{m \to \infty} \Psi(1/f_m) = \Psi(1/f)$$

which shows that $(\Psi(f))^{-1}$ belongs to the range of Ψ, as required. ∎

The following technical result will needed for the proof of Proposition V.4 below.

Lemma V.2. *Let X be a Banach space and $P : \Sigma \longrightarrow \mathcal{L}_s(X)$ be a spectral measure. If f is a P-integrable function , then*

$$\sup\{\| \int_{\Omega} g\, dP\| : g \text{ is } \Sigma\text{-measurable and } |g| \leq |f|, \; P\text{-a.e.}\} < \infty.$$

Proof. Note that each such function g is P-integrable by Corollary V.2.1. Moreover, we can write $g = f \cdot (g/f)$ where $(g/f) \in L^{\infty}(P)$ satisfies $|g/f|_P \leq 1$. Hence, by Proposition V.3 we have that

$$\int_{\Omega} g\, dP = \int_{\Omega} f \cdot (g/f)\, dP = (\int_{\Omega} f dP) \cdot (\int_{\Omega} (g/f)\, dP)$$

and so, by Theorem V.1 we get

$$\| \int_{\Omega} g\, dP\| \; \leq \; \| \int_{\Omega} (g/f)\, dP\| \cdot \| \int_{\Omega} f dP\| \leq 4|g/f|_P \cdot \|P(\Sigma)\| \cdot \| \int_{\Omega} f\, dP\|$$

$$\leq \; 4\|P(\Sigma)\| \cdot \| \int_{\Omega} f\, dP\|. \qquad \blacksquare$$

The following result shows that the *only* P-integrable functions are the P-essentially bounded functions!

Proposition V.4. *Let X be a Banach space and $P : \Sigma \longrightarrow \mathcal{L}_s(X)$ be a spectral measure . Then a Σ-measurable function $f : \Omega \longrightarrow \mathbb{C}$ is P-integrable if and only if it is P-essentially bounded.*

Proof. If f is P-essentially bounded, then f is bounded P-a.e. and so f is P-integrable; see the discussion after Definition V.2.

Conversely, suppose that $f \in L(P)$. Define *disjoint* sets

$$\delta_n(f) := \{w \in \Omega : \; n^2 \leq |f(w)| < (n+1)^2\}, \qquad n = 0, 1, 2, \ldots,$$

and Σ-simple functions

$$\phi_n(f) := \sum_{k=0}^{n} k^2 \chi_{\delta_k(f)}, \qquad n = 0, 1, 2, \ldots$$

Then $|\phi_n| \leq |f|$ pointwise on Ω, for all $n \geq 0$. By Corollary V.2.1 the function $|f|$ is P-integrable and so, by Lemma V.2, the sequence of operators

$$\int_{\Omega} \phi_n\, dP = \sum_{k=0}^{n} k^2 P(\delta_k(f)), \qquad n \geq 0,$$

is uniformly bounded. Hence, also the consecutive differences

$$\int_\Omega \phi_n \, dP - \int_\Omega \phi_{n-1} \, dP = n^2 P(\delta_n(f)), \qquad n \geq 0,$$

are uniformly bounded. So, there is $\beta > 0$ such that

$$\|P(\delta_n(f))\| \leq \beta/n^2, \qquad n \in \mathbb{N}.$$

Let $E_n := \cup_{k=n}^\infty \delta_k(f)$, for $n \geq 0$. Fix $x \in X$. The σ-additivity of P implies that $P(E_n)x = \sum_{k=n}^\infty P(\delta_k(f))x$ and so, for $n \geq 1$, we have

$$\|P(E_n)x\| \leq \sum_{k=n}^\infty \|P(\delta_k(f))x\| \leq \|x\| \cdot \sum_{k=n}^\infty \|P(\delta_k(f))\| \leq \beta\|x\| \sum_{k=n}^\infty (1/k^2).$$

Accordingly, $\|P(E_n)\| \leq \beta \sum_{k=n}^\infty (1/k^2)$, for $n \geq 1$, showing that $\|P(E_n)\| \longrightarrow 0$ as $n \to \infty$. Since $\|P(E_n)\| \geq 1$ whenever $P(E_n) \neq 0$ (see Exercise 41(a)), it follows that there exists $N > 0$ such that $P(E_n) = 0$ for all $n \geq N$. In particular, $P(E_N) = 0$ and so

$$E_N = \cup_{k=N}^\infty \delta_k(f) = \{w \in \Omega : \ N^2 \leq |f(w)|\}$$

is P-null , that is, $|f(w)| \leq N^2$ for P-a.e. $w \in \Omega$. Hence, f is P-essentially bounded. ∎

Using the results of Chapter III we are now able to identify the compact space Ω such that the Banach algebra $L^\infty(P)$ is isomorphic to $C(\Omega)$.

Theorem V.2. *Let X be a Banach space and $P : \Sigma \longrightarrow \mathcal{L}_s(X)$ be a spectral measure .*

(a) *The commutative, unital Banach algebra $L^\infty(P)$ is isomorphic to $C(\Omega)$, where Ω is the Stone space of the Bade σ-complete B.a. $P(\Sigma)$. In particular, Ω is a compact basically disconnected Hausdorff space.*

(b) *Suppose, in addition, that the spectral measure P is closed . Then $L^\infty(P)$ is Banach algebra isomorphic to $C(\Omega)$, where Ω is the Stone space of the Bade complete B.a. $P(\Sigma)$. In particular, Ω is a compact extremely disconnected Hausdorff space.*

Proof. (a) Let $\mathcal{B} := P(\Sigma)$, in which case $\mathcal{B}$ is a Bade σ-complete B.a. (see Theorem IV.1). In particular, $\mathcal{B}$ is abstractly σ-complete and so is uniformly bounded (see Theorem III.1). Let $\Omega := \Omega_\mathcal{B}$ be the Stone space of $\mathcal{B}$, in which case Ω is a compact basically disconnected Hausdorff space; see Proposition II.4. By Theorem III.2 the algebra $\langle \mathcal{B} \rangle_u^-$ is isomorphic to $C(\Omega)$. But, by Theorem V.1 the subalgebra $\Psi(L^\infty(P))$ of $\mathcal{L}_u(X)$ is isomorphic to $L^\infty(P)$, where $\Psi(f) = \int_\Omega f \, dP$, for $f \in L^\infty(P)$. So, it remains to check that $\langle \mathcal{B} \rangle_u^- = \Psi(L^\infty(P))$.

Since $\Psi(\chi_E) = P(E)$, for each $E \in \Sigma$, it is clear that $\mathcal{B} \subseteq \Psi(L^\infty(P))$ and hence, $\langle \mathcal{B} \rangle_u^- \subseteq \Psi(L^\infty(P))$. Since $\mathrm{sim}(\Sigma)$ is dense in $L^\infty(P)$, it follows that all operators which are finite sums of the form $\sum_{j \in F} \alpha_j P(E_j)$ with $\alpha_j \in \mathbb{C}$, $E_j \in \Sigma$ and $F \subseteq \mathbb{N}$ (with F finite), are dense in $\Psi(L^\infty(P))$. But, all such operators clearly belong to $\langle \mathcal{B} \rangle_u^-$ since $P(E_j) \in \mathcal{B}$ for all $j \in F$. Accordingly, $\Psi(L^\infty(P)) \subseteq \langle \mathcal{B} \rangle_u^-$.

(b) The same proof as for (a) shows that $L^\infty(P)$ is isomorphic to $C(\Omega)$, where Ω is the Stone space of $\mathcal{B} := P(\Sigma)$. But, the closedness of P is equivalent to $\mathcal{B}$ being a compact extremely disconnected space; see Theorem IV.1 and Proposition II.4. ∎

We now wish to present a refinement of Theorem III.2 in the case when the B.a. of projections $\mathcal{B}$ satisfies some additional properties.

Theorem V.3. *Let X be a Banach space and $\mathcal{B} \subseteq \mathcal{L}(X)$ be a B.a. of projections.*

(a) *Suppose that $\mathcal{B}$ is Bade σ-complete . Then the closed subalgebra $\langle \mathcal{B} \rangle_u^-$ of $\mathcal{L}_u(X)$ is isomorphic to $C(\Omega_\mathcal{B})$ as a Banach algebra (with $\Omega_\mathcal{B}$ the Stone space of $\mathcal{B}$) via an isomorphism $\Phi : C(\Omega_\mathcal{B}) \longrightarrow \langle \mathcal{B} \rangle_u^-$ given by*

$$\Phi(f) = \int_\Omega f \, d\overline{Q}, \qquad f \in C(\Omega_\mathcal{B}),$$

where $\overline{Q} : Ba(\Omega_\mathcal{B}) \longrightarrow \mathcal{L}_s(X)$ is a regular , σ-additive spectral measure satisfying $\overline{Q}(Ba(\Omega_\mathcal{B})) = \mathcal{B}$ and $\Phi(\chi_E) = \overline{Q}(E) = Q(E)$, for each $E \in Co(\Omega_\mathcal{B})$, and $Q : Co(\Omega_\mathcal{B}) \longrightarrow \mathcal{B}$ is the Stone map .

(b) *Suppose that $\mathcal{B}$ is Bade complete . Then $\langle \mathcal{B} \rangle_u^-$ is Banach algebra isomorphic to $C(\Omega_\mathcal{B})$ (with $\Omega_\mathcal{B}$ the Stone space of $\mathcal{B}$) via an isomorphism $\Phi : C(\Omega_\mathcal{B}) \longrightarrow \langle \mathcal{B} \rangle_u^-$ given by*

$$\Phi(f) = \int_\Omega f \, d\widehat{Q}, \qquad f \in C(\Omega_\mathcal{B}),$$

where $\widehat{Q} : Bo(\Omega_\mathcal{B}) \longrightarrow \mathcal{L}_s(X)$ is a regular , σ-additive, closed spectral measure satisfying $\widehat{Q}(Bo(\Omega_\mathcal{B})) = \mathcal{B}$ and $\Phi(\chi_E) = \widehat{Q}(E) = Q(E)$, for each $E \in Co(\Omega_\mathcal{B})$, and $Q : Co(\Omega_\mathcal{B}) \longrightarrow \mathcal{B}$ is the Stone map.

Proof. (a) By Theorem III.1 $\mathcal{B}$ is uniformly bounded and by Theorem III.2 the algebra $\langle \mathcal{B} \rangle_u^-$ is isomorphic to $C(\Omega_\mathcal{B})$. By Theorem III.3 (with $\Omega_\mathcal{B}$ in place of Λ) there is a unique bounded spectral measure $R : Bo(\Omega_\mathcal{B}) \longrightarrow \mathcal{L}(X')$ of class $\Gamma = X \subseteq X''$ such that $\langle x, Rx' \rangle : Bo(\Omega_\mathcal{B}) \longrightarrow \mathbb{C}$ is regular for all $x \in X$ and $x' \in X'$ and, for each $f \in C(\Omega_\mathcal{B})$, we have

$$(11) \qquad \langle \Phi(f)x, x' \rangle = \int_{\Omega_\mathcal{B}} f \, d\langle x, Rx' \rangle, \qquad x \in X, \, x' \in X',$$

where $\Phi : C(\Omega_\mathcal{B}) \longrightarrow \langle \mathcal{B} \rangle_u^-$ is the isomorphism of Theorem III.2.

Let Σ denote the family of those Borel sets $E \subseteq \Omega_\mathcal{B}$ for which $R(E) = \overline{Q}(E)'$ for some projection $\overline{Q}(E) \in \mathcal{B}$. Since $\mathcal{B}$ is a B.a. it follows that Σ is an *algebra of subsets* of $\Omega_\mathcal{B}$ (contained in $Bo(\Omega_\mathcal{B})$). To see that Σ is a σ-algebra, let $\{E_n\}_{n=1}^\infty$ be an increasing sequence of sets from Σ. Since

$$\overline{Q}(E_{n+1})' \, \overline{Q}(E_n)' = R(E_{n+1}) R(E_n) = R(E_{n+1} \cap E_n) = R(E_n) = \overline{Q}(E_n)'$$

we see that

$$\langle \overline{Q}(E_n) \overline{Q}(E_{n+1}) x, x' \rangle = \langle x, \overline{Q}(E_{n+1})' \, \overline{Q}(E_n)' x' \rangle = \langle x, \overline{Q}(E_n)' x' \rangle = \langle Q(E_n)x, x' \rangle,$$

for all $n \in \mathbb{N}$, $x \in X$, $x' \in X'$ which implies that $\overline{Q}(E_n) \overline{Q}(E_{n+1}) = \overline{Q}(E_n)$, for all n. Hence, $\{\overline{Q}(E_n)\}_{n=1}^\infty$ is increasing in $\mathcal{B}$. By Theorem IV.1 and Lemma IV.1 it follows that

$B := \lim_{n \to \infty} \overline{Q}(E_n)$ exists in $\mathcal{L}_s(X)$ and $B \in \mathcal{B}$. Thus, for each $x \in X$ and $x' \in X'$ we have, by σ-additivity of $\langle x, Rx' \rangle$, that

$$\langle x, R(\cup_{n=1}^\infty E_n) x' \rangle = \lim_{n \to \infty} \langle x, R(E_n) x' \rangle = \lim_{n \to \infty} \langle \overline{Q}(E_n) x, x' \rangle = \langle Bx, x' \rangle = \langle x, B'x' \rangle.$$

This shows that $\cup_{n=1}^\infty E_n \in \Sigma$ and hence, that Σ is a σ-*algebra* .

The isomorphism Φ satisfies $\Phi(\chi_E) = Q(E)$, for each $E \in Co(\Omega_\mathcal{B})$; see Theorem III.2. Since $\Phi(\chi_E) = Q(E)$ is a projection in $\mathcal{L}(X)$ and satisfies $Q(E)' = R(E)$ by the formula (11), it follows that $\overline{Q}(E) = Q(E)$ and $Co(\Omega_\mathcal{B}) \subseteq \Sigma$. Hence, also $Ba(\Omega_\mathcal{B}) \subseteq \Sigma$ (see Proposition II.5). Since $\langle \overline{Q}(E)x, x' \rangle = \langle x, R(E)x' \rangle$ for each $x \in X$, $x' \in X'$ and $E \in Ba(\Omega_\mathcal{B})$ it is clear that $\overline{Q} : Ba(\Omega_\mathcal{B}) \longrightarrow \mathcal{L}_s(X)$ is σ-additive, regular and multiplicative (as R is multiplicative). That is, $\overline{Q}$ is a regular, σ-additive spectral measure . Since $\overline{Q}(Co(\Omega_\mathcal{B})) = \mathcal{B}$ it follows from the definition of Σ that $\overline{Q}(Ba(\Omega_\mathcal{B})) = \mathcal{B}$.

(b) If $\mathcal{B}$ is Bade complete , then the same argument establishes that Σ is a σ-algebra . It remains to establish that $\Sigma = Bo(\Omega_\mathcal{B})$. Since $\mathcal{B}$ has the monotone property (see Theorem IV.1) it follows from Theorem IV.2(b) that there exists a unique *closed* spectral measure $\widehat{Q} : Bo(\Omega_\mathcal{B}) \longrightarrow \mathcal{L}_s(X)$ such that $\widehat{Q}(Bo(\Omega_\mathcal{B})) = \mathcal{B}$ and $\widehat{Q}(E) = Q(E)$ for all $E \in Co(\Omega_\mathcal{B})$, and that $\widehat{Q}$ is *regular* (see also the proof of Theorem IV.2(b)). In particular, $\langle \widehat{Q}x, x' \rangle$ is a regular complex measure on $Bo(\Omega_\mathcal{B})$, for each $x \in X$ and $x' \in X'$, and coincides with the regular measure $\langle x, Rx' \rangle$ on the algebra $Co(\Omega_\mathcal{B})$ and hence, also on the generated σ-algebra $\sigma(Co(\Omega_\mathcal{B})) = Ba(\Omega_\mathcal{B})$; see Exercise 36. Since the regular extension to the Borel sets of any regular Baire measure on a compact Hausdorff space is *unique* (see [37; p.314]) it follows that $\langle \widehat{Q}x, x' \rangle = \langle x, Rx' \rangle$ as measures on $Bo(\Omega_\mathcal{B})$. In particular,

$$\langle x, \widehat{Q}(E)'x' \rangle = \langle \widehat{Q}(E)x, x' \rangle = \langle x, R(E)x' \rangle, \qquad E \in Bo(\Omega_\mathcal{B}),$$

for each $x \in X$ and $x' \in X'$, from which it follows that $\widehat{Q}(E)' = R(E)$ for all $E \in Bo(\Omega_\mathcal{B})$. Hence, $Bo(\Omega_\mathcal{B}) \subseteq \Sigma$ and so $Bo(\Omega_\mathcal{B}) = \Sigma$.

Since $\widehat{Q}(Co(\Omega_\mathcal{B})) = \mathcal{B}$ it follows from the definition of Σ that $\widehat{Q}(Bo(\Omega_\mathcal{B})) = \mathcal{B}$.   ∎

Remark.   (a) Theorem V.3. should be compared with [15; Lemma 9, p.2202]. The proof given there is in the spirit of Banach algebra theory, whereas our proof is based on the theory of B.a. 's and Stone spaces . In particular, our proof that $Bo(\Omega_\mathcal{B})$ equals Σ (in part (b) of Theorem V.3) is quite different to the argument given in [15].

(b) It is clear from the proof of Theorem V.3 that the spectral measures $\overline{Q}$ and $\widehat{Q}$ are precisely the more general σ-homomorphisms of Theorem II.4 and Theorem II.5 specialized to the particular setting of B.a. 's of *projections* (which are Bade σ-complete and Bade complete).

(c) Let $\mathcal{B}$ be a bounded B.a. of projections and $Q : Co(\Omega_\mathcal{B}) \longrightarrow \mathcal{B}$ be the Stone map . Then we have seen (at the end of Chapter III) that each regular, bounded, σ-additive measure $\langle Qx, x' \rangle$, which is defined on the *algebra of sets* $Co(\Omega_\mathcal{B})$ for each $x \in X$ and $x' \in X'$, has an extension to a regular , σ-additive measure $\mu_{x,x'} : Ba(\Omega_\mathcal{B}) \longrightarrow \mathbb{C}$. By Theorem III.3 there also exists a regular, σ-additive measure $\langle x, Rx' \rangle : Bo(\Omega_\mathcal{B}) \longrightarrow \mathbb{C}$ which coincides with

$\mu_{x,x'}$ on $Ba(\Omega_B)$. The question raised at the end of Chapter III was to decide when each projection from the B.a. $R(Ba(\Omega_B))$ or $R(Bo(\Omega_B))$, rather than just those from $R(Co(\Omega_B))$, is the dual operator of some projection from $\mathcal{B}$? An examination of the proof of Theorem V.3 provides an answer for certain $\mathcal{B}$. Namely, if $\mathcal{B}$ is Bade σ-complete (resp. Bade complete), then each projection from $R(Ba(\Omega_B))$ (resp. $R(Bo(\Omega_B))$) is indeed of the form B' for some $B \in \mathcal{B}$. The following exercise shows that this *fails* if the Bade σ-completeness or Bade completeness assumption of $\mathcal{B}$ is reduced merely to its abstract σ-completeness or abstract completeness. In particular, it fails for bounded B.a.'s of projections in general. ∎

Exercise 45. Let $X = \ell^\infty$ and let S denote the σ-algebra of all subsets of N. For each $E \in S$ define $B(E) \in \mathcal{L}(X)$ by

$$B(E)x = (\chi_E(1)x_1, \chi_E(2)x_2, \dots), \qquad x = (x_1, x_2, \dots) \in X.$$

It is shown in Example 20(a) that the B.a. $\mathcal{B} := \{B(E) : E \in S\}$ is abstractly complete, but *not* Bade σ-complete. Observe that $\|\mathcal{B}\| = \sup\{\|B(E)\| : E \in S\} = 1$. Let Ω_B be the Stone space of $\mathcal{B}$ and $\Phi : C(\Omega_B) \longrightarrow \langle \mathcal{B} \rangle_u^-$ be the Banach algebra isomorphism of Theorem III.2. By Theorem III.3 there is a unique bounded spectral measure $R : Bo(\Omega_B) \longrightarrow \mathcal{L}(X')$ of class $\Gamma = X \subseteq X''$ such that $\langle x, Rx' \rangle$ is a regular measure on $Bo(\Omega_B)$, for each $x \in X$ and $x' \in X'$, and for each $f \in C(\Omega_B)$ we have

$$\langle \Phi(f)x, x' \rangle = \int_{\Omega_B} f \, d\langle x, Rx' \rangle, \qquad x \in X, \, x' \in X'.$$

Let Σ denote the family of those sets $E \in Bo(\Omega_B)$ for which $R(E) = P(E)'$ for some projection $P(E) \in \mathcal{L}(X)$.

(a) Show that Σ is an algebra of sets containing $Co(\Omega_B)$.

(b) Show that Σ is *not* a σ-algebra. ∎

So far in this chapter we have seen that if $P : \Sigma \longrightarrow \mathcal{L}_s(X)$ is a σ-additive spectral measure, then the space $L(P)$ of all P-integrable functions consists precisely of the P-essentially bounded functions. Moreover, if we equip (the space of equivalence classes of) $L(P)$ with the *norm* $|\cdot|_P$, then we produce the Banach space (and algebra) $L^\infty(P)$ and the map $f \mapsto \int_\Omega f \, dP$ is a bicontinuous isomorphism of $L^\infty(P)$ onto $\langle \mathcal{B} \rangle_u^- \simeq C(\Omega_B)$, where $\mathcal{B} := P(\Sigma)$. For the remainder of this chapter we wish to concentrate on identifying the closed algebra generated by $\mathcal{B}$ in $\mathcal{L}_s(X)$ or, equivalently, in $\mathcal{L}_w(X)$. For this purpose we will need to consider (the space of equivalence classes of) $L(P)$ equipped with a locally convex Hausdorff topology which is *non-normable* and to investigate, in detail, the algebraic and continuity properties of the linear map $f \mapsto \int_\Omega f \, dP$ with respect to this locally convex topology (rather than the norm topology $|\cdot|_P$). The results on vector measures developed in Chapter I will play a crucial role in this respect.

Given $f \in L(P)$, where $P : \Sigma \longrightarrow \mathcal{L}_s(X)$ is a spectral measure, we define the equivalence class $[f] = \{g \in L(P) : |f - g| \text{ is } P\text{-null}\}$ and, as usual, let $L^1(P) = \{[f] : f \in L(P)\}$. Then we have seen that $L^1(P) = L^\infty(P)$ as a vector space and hence, $L^1(P)$ becomes a Banach algebra with respect to the norm $|\cdot|_P$. We now equip $L^1(P)$ with a different topology. By the usual abuse of notation we still denote elements of $L^1(P)$ by f, rather than $[f]$.

For each fixed $x \in X$, define a *seminorm* $q_x(P) : L^1(P) \longrightarrow [0, \infty)$ by

$$q_x(P)(f) := \|f\|_{L^1(Px)} = \|(Px)_f\|(\Omega), \qquad f \in L^1(P),$$

where we recall that $Px : \Sigma \longrightarrow X$ is the vector measure $E \mapsto P(E)x$, for $E \in \Sigma$, and $(Px)_f : \Sigma \longrightarrow X$ is the vector measure given by the indefinite integral of f with respect to Px, that is, by $E \mapsto \int_E f\, d(Px)$, for $E \in \Sigma$. Of course, $\|(Px)_f\|(\cdot)$ denotes the semivariation of $(Px)_f$; see Chapter I. Observe that Proposition V.2 ensures $f \in L^1(Px)$ whenever $f \in L^1(P)$. The topology $\tau_s(P)$ specified by the family of seminorms $\{q_x(P) : x \in X\}$ turns $L^1(P)$ into a locally convex space. Moreover, the topology $\tau_s(P)$ is *Hausdorff* meaning that if $f \in L^1(P)$ satisfies $q_x(P)(f) = 0$ for all $x \in X$, then $f = 0$ in $L^1(P)$. To see this fix $E \in \Sigma$. For each $x \in X$, it follows from Proposition I.2 applied to the vector measure $m = (Px)_f$ that $\int_E f\, dP \in \mathcal{L}(X)$ satisfies

$$
\begin{aligned}
\|(\int_E f\, dP)x\| &= \|\int_E f\, d(Px)\| = \|(Px)_f(E)\| \leq \|(Px)_f\|(E) \\
&\leq \|(Px)_f\|(\Omega) = q_x(P)(f) = 0.
\end{aligned}
$$

Since $x \in X$ is arbitrary, it follows that $\int_E f\, dP = 0$ (in $\mathcal{L}(X)$). Since this is true for every $E \in \Sigma$, it follows that $f = 0$ in $L^1(P)$. So, we see that $(L^1(P), \tau_s(P))$ is a locally convex Hausdorff space. Proposition V.3 shows that $L^1(P)$ is an algebra; the next result shows that $L^1(P)$ is actually a *locally convex algebra*. This means that multiplication is *separately continuous*, that is, if $\{f_\alpha\} \subseteq L^1(P)$ is a net such that $f_\alpha \longrightarrow 0$ in $L^1(P)$, then for each fixed $g \in L^1(P)$, also $gf_\alpha \stackrel{\alpha}{\longrightarrow} 0$ and $f_\alpha g \stackrel{\alpha}{\longrightarrow} 0$ in $L^1(P)$. Of course, since $L^1(P)$ is a *commutative* algebra it suffices to check that either one of $gf_\alpha \stackrel{\alpha}{\longrightarrow} 0$ or $f_\alpha g \stackrel{\alpha}{\longrightarrow} 0$ is satisfied.

Definition V.3. Let $P : \Sigma \longrightarrow \mathcal{L}_s(X)$ be a spectral measure . Then the *integration map* $I_P : L^1(P) \longrightarrow \mathcal{L}(X)$ is defined by

$$I_P(f) := P(f) = \int_\Omega f\, dP, \qquad f \in L^1(P). \qquad \blacksquare$$

Theorem V.4. *Let X be a Banach space and $P : \Sigma \longrightarrow \mathcal{L}_s(X)$ be a spectral measure . Then $(L^1(P), \tau_s(P))$ is a locally convex (commutative) algebra with identity $\mathbb{1}$ and*

(12) $$q_x(I_P(f)) \leq q_x(P)(f) \leq 4\|P(\Sigma)\| \cdot q_x(I_P(f)), \qquad f \in L^1(P),$$

for each $x \in X$, where $q_x : \mathcal{L}_s(X) \longrightarrow [0, \infty)$ is the continuous seminorm defined by $q_x(T) := \|Tx\|$, for each $T \in \mathcal{L}(X)$. In particular, the integration map I_P is a bicontinuous (algebra) isomorphism of $L^1(P)$ onto its range $I_P(L^1(P))$ equipped with the relative topology from $\mathcal{L}_s(X)$.

Proof. We have already seen that $L^1(P)$ is a commutative algebra with unit the constant function $\mathbb{1}$ on Ω. Let us establish (12). So, fix $x \in X$. Then, by Proposition I.2 applied to

$m := (Px)_f$, we have

(13) $\quad q_x(I_P(f)) = \|P(f)x\| = \|\int_\Omega f\, d(Px)\| = \|(Px)_f(\Omega)\| \le \|(Px)_f\|(\Omega) = q_x(P)(f)$,

for each $f \in L^1(P)$. But, again by Proposition I.2, for $f \in L^1(P)$ we have

(14) $\quad q_x(P)(f) = \|(Px)_f\|(\Omega) \le 4 \sup\{\|(Px)_f(E)\| : E \in \Sigma\}$

$$= 4 \sup\{\|\int_E f\, dPx\| : E \in \Sigma\} = 4 \sup\{\|P(E)P(f)x\| : E \in \Sigma\},$$

where the last equality follows from the identity (3) and Proposition V.1. But, the inequality $\|P(E)P(f)x\| \le \|P(E)\| \cdot \|P(f)x\|$, valid for each $E \in \Sigma$, implies that

$$\sup\{\|P(E)P(f)x\| : E \in \Sigma\} \le \|P(\Sigma)\| \cdot \|P(f)x\| = \|P(\Sigma)\| \cdot q_x(I_P(f)).$$

Combining this inequality with (13) and (14) yields (12).

Since I_P is injective (Proposition V.1 shows that $\int_E f\, dP = 0$ for all $E \in \Sigma$, that is, $f = 0$ in $L^1(P)$, whenever $P(f) = I_P(f) = 0$), it follows from (12) that I_P is a bicontinuous isomorphism of $L^1(P)$ onto its range in $\mathcal{L}_s(X)$, recalling that the seminorms $\{q_x : x \in X\}$ determine the topology of $\mathcal{L}_s(X)$.

Finally, it follows from (12) that, for fixed $x \in X$, we have

$$(4\|P(\Sigma)\|)^{-1} q_x(P)(fg) \le q_x(I_P(fg)) = q_x(I_P(f)I_P(g)) = \|I_P(f)I_P(g)x\|$$
$$\le \|I_P(g)\| \cdot \|I_P(f)x\| = \|I_P(g)\| \cdot q_x(I_P(f)) \le \|I_P(g)\| \cdot q_x(P)(f),$$

for all f and g in $L^1(P)$, that is,

$$q_x(P)(fg) \le 4\|P(\Sigma)\| \cdot \|I_P(g)\| \cdot q_x(P)(f).$$

It follows if $g \in L^1(P)$ is *fixed* and $f_\alpha \longrightarrow 0$ in $L^1(P)$, then also $gf_\alpha \longrightarrow 0$ in $L^1(P)$. Hence $(L^1(P), \tau_s(P))$ is a locally convex algebra. ∎

Exercise 46. Let X be a Banach space and $P : \Sigma \longrightarrow \mathcal{L}_s(X)$ be a spectral measure.

(a) Show that the simple functions $\text{sim}(\Sigma)$ are sequentially dense in $L^1(P)$ for the topology $\tau_s(P)$.

(b) Show that the identity function from the Banach space $(L^\infty(P), |\cdot|_P)$ into the locally convex Hausdorff space $(L^1(P), \tau_s(P))$ is continuous. ∎

In view of Theorem IV.1 it is clear that *closed* spectral measures will play an important role in determining the closed algebra in $\mathcal{L}_s(X)$ generated by a *Bade complete* B.a. of projections. The following result, which will require Theorem V.4, gives some useful characterizations of closed spectral measures.

Theorem V.5. *Let X be a Banach space and $P : \Sigma \longrightarrow \mathcal{L}_s(X)$ be a spectral measure. Then the following statements are equivalent.*

(a) *The range $P(\Sigma) := \{P(E) : E \in \Sigma\}$ is a closed subset of $\mathcal{L}_s(X)$.*

(b) *The range $P(\Sigma)$ is a complete subset of the locally convex Hausdorff space $\mathcal{L}_s(X)$.*

(c) *The range $P(\Sigma)$ is a Bade complete B.a. of projections.*

(d) *P is a closed spectral measure, that is, the uniform space $(\Sigma(P), \tau_s(P))$ is complete.*

Proof. (b)$\Rightarrow$(a). This follows from the fact that any complete subset of a topological space is necessarily a closed set.

(a)$\Rightarrow$(b). Let $\{B_\alpha\}_{\alpha \in A}$ be a net from $P(\Sigma)$ which is Cauchy in $\mathcal{L}_s(X)$. Fix $x \in X$. Then

$$\|B_\alpha x - B_\beta x\| = q_x(B_\alpha - B_\beta), \qquad \alpha, \beta \in A,$$

and so $\{B_\alpha x\}_{\alpha \in A}$ is Cauchy in X. By the completeness of X there is $Bx \in X$ such that $\lim_\alpha B_\alpha x = Bx$. Since $\sup_\alpha \|B_\alpha\| \leq \|P(\Sigma)\| < \infty$ (c.f. Lemma III.3) it follows from the Banach-Steinhaus theorem (c.f. Proposition III.1(b)) that $B \in \mathcal{L}(X)$. Accordingly, $B_\alpha \longrightarrow B$ in $\mathcal{L}_s(X)$. Since $P(\Sigma)$ is assumed to be a closed set, we deduce that $B \in P(\Sigma)$. Hence, $P(\Sigma)$ is complete .

(d)$\Rightarrow$(c). This is part of Theorem IV.1.

(c)$\Rightarrow$(d). By Definition IV.1(i) of Bade completeness the B.a. $P(\Sigma)$ is abstractly complete . Since $\Sigma(P)$ is B.a. isomorphic to $P(\Sigma)$, also $\Sigma(P)$ is abstractly complete. Suppose that $\{[E_\alpha]\} \subseteq (\Sigma(P), \tau_s(P))$ is a net which is downwards filtering to $[\emptyset]$. Then $\{P(E_\alpha)\} \subseteq P(\Sigma)$ is a net which is decreasing to 0 (in the order of $P(\Sigma)$). Since $P(\Sigma)$ has the ordered convergence property (c.f. Theorem IV.1) it follows from Lemma IV.1.(b) that $\lim_\alpha P(E_\alpha) = \wedge_\alpha P(E_\alpha) = 0$ in $\mathcal{L}_s(X)$. Then Lemma IV.3 implies that P is a closed spectral measure.

(b)$\Rightarrow$(d). Suppose that $P(\Sigma)$ is a complete subset of $\mathcal{L}_s(X)$. Let $\{[E_\alpha]\}$ be a Cauchy net in $(\Sigma(P), \tau_s(P))$. Fix $x \in X$. Then

$$(15) \qquad q_x(P(E_\alpha) - P(E_\beta)) = \|Px(E_\alpha) - Px(E_\beta)\| = \left\| \int_\Omega (\chi_{E_\alpha} - \chi_{E_\beta}) \, dPx \right\|$$

$$= q_x(I_P(\chi_{E_\alpha} - \chi_{E_\beta})) \leq q_x(P)(\chi_{E_\alpha} - \chi_{E_\beta}),$$

for all α and β, where the last inequality follows from (12). But $q_x(P)(\chi_{E_\alpha} - \chi_{E_\beta}) = q_x(P)(|\chi_{E_\alpha} - \chi_{E_\beta}|)$, by the Remark after Proposition I.5 applied to $m := Px$ and $f := \chi_{E_\alpha} - \chi_{E_\beta}$, and the identity $|\chi_E - \chi_F| = \chi_{E \triangle F}$ for all $E, F \in \Sigma$. So, (15) implies that

$$(16) \qquad q_x(P(E_\alpha) - P(E_\beta)) \leq q_x(P)(\chi_{E_\alpha \triangle E_\beta}) = \|Px\|(E_\alpha \triangle E_\beta),$$

for all α and β. By definition of the topology and uniform structure in $\Sigma(P)$, as defined by the pseudometrics $\{d_x : x \in X\}$ specified after Exercise 37 in Chapter IV, (16) implies that $\{P(E_\alpha)\}$ is a Cauchy net in $P(\Sigma)$. Then the topological completeness of $P(\Sigma)$ in $\mathcal{L}_s(X)$ ensures there is $E \in \Sigma$ such that $P(E_\alpha) \longrightarrow P(E)$ in $\mathcal{L}_s(X)$. So, for each α (with f denoting $(\chi_E - \chi_{E_\alpha})$), we have by (12) that

$$\begin{aligned} d_x([E], [E_\alpha]) &= \|Px\|(E \triangle E_\alpha) = \|(Px)_{|f|}\|(\Omega) = \|(Px)_f\|(\Omega) \\ &= q_x(P)(f) \leq 4\|P(\Sigma)\| \cdot q_x(I_P(f)) = 4\|P(\Sigma)\| \cdot q_x(P(E) - P(E_\alpha)). \end{aligned}$$

Since $P(E_\alpha) \longrightarrow P(E)$ in $\mathcal{L}_s(X)$ it follows that $[E_\alpha] \overset{\alpha}{\longrightarrow} [E]$ in $\Sigma(P)$. Accordingly, $\Sigma(P)$ is $\tau_s(P)$ complete , that is, P is closed spectral measure .

(d)$\Rightarrow$(b). Suppose that P is a closed spectral measure . Let $\{P(E_\alpha)\}_{\alpha \in A}$ be a net which is Cauchy in $P(\Sigma)$ for the relative topology from $\mathcal{L}_s(X)$. Fix $x \in X$. We saw in the proof of (b)$\Rightarrow$(d) that

$$\|Px\|(E_\alpha \triangle E_\beta) \leq 4\|P(\Sigma)\| \cdot q_x(P(E_\alpha) - P(E_\beta)), \qquad \alpha, \beta \in A,$$

and hence , $\{[E_\alpha]\}_{\alpha \in A}$ is Cauchy in $(\Sigma(P), \tau_s(P))$. So, there exists $[E] \in \Sigma(P)$ such that $[E_\alpha] \overset{\alpha}{\longrightarrow} [E]$ in $\Sigma(P)$. The inequality

$$q_x(P(E_\alpha) - P(E)) \leq \|Px\|(E \triangle E_\alpha) = d_x([E], [E_\alpha]), \qquad \alpha \in A,$$

was also established in the proof of (b)$\Rightarrow$(d) and hence, we see that $P(E_\alpha) \longrightarrow P(E) \in P(\Sigma)$ in $\mathcal{L}_s(X)$. This shows that $P(\Sigma)$ is topologically complete . ∎

The following exercise shows that for B.a.'s of projections $\mathcal{B}$ (even abstractly complete ones) which are *not* the range of some spectral measure, the properties (a)–(d) of Theorem V.5 are no longer equivalent, in general.

Exercise 47. Let $X = \ell^\infty$ and $\Sigma = 2^{\mathbb{N}}$. For each $E \in \Sigma$ define $B(E) \in \mathcal{L}(X)$ by

$$B(E)x = (\chi_E(1)x_1, \chi_E(2)x_2, \dots), \qquad x = (x_1, x_2, \dots) \in \ell^\infty,$$

in which case $\mathcal{B} := \{B(E) : E \in \Sigma\}$ is an abstractly complete B.a. of projections which is not Bade σ-complete ; see Example 20(a). So, $\mathcal{B}$ cannot be the range of any spectral measure; see Theorem IV.1. Nevertheless, show that $\mathcal{B}$ is a *closed subset* of $\mathcal{L}_s(X)$. ∎

The next result is a useful consequence of Theorem V.5.

Corollary V.5.1. *Every Bade complete B.a. of projections in $\mathcal{L}(X)$, with X a Banach space, is a closed (= complete) subset of $\mathcal{L}_s(X)$.*

Proof. If $\mathcal{B} \subseteq \mathcal{L}(X)$ is a Bade complete B.a. , then Theorem IV.1 implies that $\mathcal{B}$ coincides with the range of some *closed* spectral measure . The desired conclusion then follows from Theorem V.5. ∎

We are finally in the position where we can describe the closed subalgebra in $\mathcal{L}_s(X)$ generated by Bade σ-complete and Bade complete B.a.'s of projections. The following result will be crucial in this regard. So, let $\mathcal{B}$ be a B.a. of projections from $\mathcal{L}(X)$. Then the smallest closed subalgebra of $\mathcal{L}_s(X)$ (resp. $\mathcal{L}_w(X)$) generated by $\mathcal{B}$ is denoted by $\langle \mathcal{B} \rangle_s^-$ (resp. $\langle \mathcal{B} \rangle_w^-$). Since the linear span of $\mathcal{B}$ in $\mathcal{L}(X)$ is a *convex* set, it follows from Proposition III.1(d) that $\langle \mathcal{B} \rangle_s^- = \langle \mathcal{B} \rangle_w^-$ as vector subspaces of $\mathcal{L}(X)$.

Theorem V.6. *Let X be a Banach space and $P : \Sigma \longrightarrow \mathcal{L}_s(X)$ be a closed spectral measure.*
  (a) *The locally convex Hausdorff space $(L^1(P), \tau_s(P))$ is complete .*
  (b) *The integration map $I_P : L^1(P) \longrightarrow \mathcal{L}_s(X)$ is a bicontinuous isomorphism of $L^1(P)$ onto $\langle P(\Sigma) \rangle_s^-$.*

Proof. It was noted in the proof of Lemma IV.3 that the locally convex Hausdorff space $\mathcal{L}_s(X)$ is quasicomplete . Let Z denote the *completion* of $\mathcal{L}_s(X)$ and $\widetilde{P} : \Sigma \longrightarrow Z$ denote P

considered as being Z-valued, in which case $\widetilde{P}$ is also σ-additive. We note that the L^1-space of a vector measure with values in a locally convex Hausdorff space is defined analogously as for Banach spaces; see [19], [27], for example. It is clear that $L^1(P) \subseteq L^1(\widetilde{P})$ (as a vector space inclusion) and that $\tau(\widetilde{P})$ induces the topology $\tau(P)$ on $L^1(P)$. To see that actually $L^1(P) = L^1(\widetilde{P})$ it suffices to show that each $\widetilde{P}$-integrable function $f \geq 0$ is P-integrable. Choose Σ-simple functions $s_n \geq 0$, $n \in \mathbb{N}$, such that $s_n \uparrow f$ pointwise. Since $\int_E s_n d\widetilde{P} \in \mathcal{L}_s(X)$, for each $E \in \Sigma$ and $n \in \mathbb{N}$, it follows from the dominated convergence theorem for general vector measures (see [27; Chapter II] or [29; Theorem 2.2], for example) and the quasicompleteness of $\mathcal{L}_s(X)$ that $\int_E f d\widetilde{P} = \lim_{n\to\infty} \int_E s_n d\widetilde{P}$ belongs to $\mathcal{L}_s(X)$. Hence, f is P-integrable with $\int_E f dP = \int_E f d\widetilde{P}$ for each $E \in \Sigma$. Accordingly, $L^1(P) = L^1(\widetilde{P})$ with equality as locally convex spaces. In particular, since $\Sigma(P)$ (resp. $\Sigma(\widetilde{P})$) is identified with the subset $\{\chi_E : E \in \Sigma\}$ of $L^1(P)$ (resp. $L^1(\widetilde{P})$) it follows that $\widetilde{P}$ is a closed measure because P is closed (by hypothesis). But, the completeness of Z then implies that $L^1(\widetilde{P})$ is a complete locally convex Hausdorff space, [27; Theorem 1, p.73], and hence, $L^1(P)$ is also complete (being equal to $L^1(\widetilde{P})$). This establishes part (a).

By Theorem V.4 the integration map I_P is a bicontinuous isomorphism of $L^1(P)$ onto its range $I_P(L^1(P)) \subseteq \mathcal{L}_s(X)$. So, it remains to check that $\langle P(\Sigma) \rangle_s^- = I_P(L^1(P))$.

Since $P(E) = I_P(\chi_E)$ for each $E \in \Sigma$, it is clear that $P(\Sigma) \subseteq I_P(L^1(P))$. Since $I_P(L^1(P))$ is complete, it is also closed in $\mathcal{L}_s(X)$ and so $\langle P(\Sigma) \rangle_s^- \subseteq I_P(L^1(P))$. Conversely, it is clear that $I_P(\mathrm{sim}(\Sigma)) \subseteq \langle P(\Sigma) \rangle_s^-$. Since $\mathrm{sim}(\Sigma)$ is $\tau_s(P)$ dense in $L^1(P)$ by Exercise 46, and the integration map I_P is continuous, it follows that $I_P(L^1(P)) \subseteq \langle P(\Sigma) \rangle_s^-$. ∎

Remark.   If a Banach space X is infinite dimensional, then the quasicomplete space $\mathcal{L}_s(X)$ is *never* complete , that is, there always exist Cauchy nets with no limit in $\mathcal{L}_s(X)$. However, the closed subalgebra $\langle P(\Sigma) \rangle_s^- \subseteq \mathcal{L}_s(X)$ is always complete in the case when P is a closed spectral measure. ∎

We can now combine the results of this chapter with those of Chapter IV to deduce some interesting facts about Bade complete B.a. 's of projections.

Corollary V.6.1.   *Let X be a Banach space and $\mathcal{B} \subseteq \mathcal{L}_s(X)$ be a Bade complete B.a. of projections. If $B \in \mathcal{L}(X)$ is a projection such that $B \in \langle \mathcal{B} \rangle_w^-$, then necessarily $B \in \mathcal{B}$.*

Proof.   Let $\Omega_\mathcal{B}$ denote the Stone space of $\mathcal{B}$ in which case we know that there exists a *closed* spectral measure $P : Bo(\Omega_\mathcal{B}) \longrightarrow \mathcal{L}_s(X)$ such that $P(Bo(\Omega_\mathcal{B})) = \mathcal{B}$; see Theorem IV.1 and Theorem IV.2. Since $\langle \mathcal{B} \rangle_w^- = \langle \mathcal{B} \rangle_s^-$ and $I_P : L^1(P) \longrightarrow \mathcal{L}_s(X)$ is an isomorphism onto $\langle \mathcal{B} \rangle_s^-$ (c.f. Theorem V.6), there exists $f \in L^1(P)$ such that $I_P(f) = B$. As I_P is a homomorphism, we deduce that

$$I_P(f^2 - f) = [I_P(f)]^2 - I_P(f) = B^2 - B = 0$$

and so $f^2 = f$ (as I_P is injective) . That is, $g = f^2 - f$ is a P-null function and so $E := \{w \in \Omega : g(w) \neq 0\}$ is a P-null set . Since $\Omega \backslash E$ is the disjoint union of the sets $E_0 := \{w : f(w) = 0\}$ and $E_1 := \{w : f(w) = 1\}$, and E is P-null, it follows from the identity

$$B = P(f) = P(f)[P(E) + P(E_0) + P(E_1)]$$

that $B = P(f)P(E_1) = \int_{E_1} f dP = P(E_1)$. Hence, $B \in \mathcal{B}$. ∎

Exercise 48. Give an example of a Bade σ-complete B.a. of projections $\mathcal{B} \subseteq \mathcal{L}(X)$ for which the conclusion of Corollary V.6.1 *fails* to hold. *Hint*: Consider Example 20(b). ∎

Combining Theorem V.6 and the proof of Corollary V.6.1 yields the following important representation theorem.

Corollary V.6.2. *Let X be a Banach space and $\mathcal{B} \subseteq \mathcal{L}(X)$ be a Bade complete B.a. of projections. Then $\langle \mathcal{B} \rangle_s^-$ is isomorphic as a locally convex algebra to $L^1(P)$, where $P :$ $Bo(\Omega_\mathcal{B}) \longrightarrow \mathcal{B} \subseteq \mathcal{L}_s(X)$ is a regular, closed spectral measure which extends the Stone map $Q : Co(\Omega_\mathcal{B}) \longrightarrow \mathcal{B}$ and $\Omega_\mathcal{B}$ is the Stone space of $\mathcal{B}$.*

A further consequence is the following interesting fact.

Corollary V.6.3. *Let $\mathcal{B} \subseteq \mathcal{L}(X)$ be a Bade complete B.a. of projections. Then*

$$\langle \mathcal{B} \rangle_u^- = \langle \mathcal{B} \rangle_s^- = \langle \mathcal{B} \rangle_w^-.$$

Moreover, $\langle \mathcal{B} \rangle_s^-$ is an inverse closed subalgebra of $\mathcal{L}(X)$.

Proof. It was already noted that the equality $\langle \mathcal{B} \rangle_s^- = \langle \mathcal{B} \rangle_w^-$ is always valid. Since $\mathcal{L}_u(X)$ is continuously included in $\mathcal{L}_s(X)$, because

$$q_x(T) := \|Tx\| \leq \|T\|_{\mathcal{L}_u(X)} \cdot \|x\|, \qquad T \in \mathcal{L}(X),$$

for each $x \in X$, it is clear that $\langle \mathcal{B} \rangle_u^- \subseteq \langle \mathcal{B} \rangle_s^-$. Conversely, if $T \in \langle \mathcal{B} \rangle_s^-$, then $T = I_P(f)$ for some $f \in L^1(P)$, where $P : Bo(\Omega_\mathcal{B}) \longrightarrow \mathcal{L}_s(X)$ is a closed spectral measure as in Corollary V.6.2. By Proposition V.4, $f \in L^\infty(P)$ and so we can choose $Bo(\Omega_\mathcal{B})$-simple functions $\{s_n\}_{n=1}^\infty$ such that $s_n \longrightarrow f$ in $L^\infty(P)$. Theorem V.1 implies that $I_P(s_n) \longrightarrow I_P(f)$ in $\mathcal{L}_u(X)$. Since each operator $I_P(s_n) \in \langle \mathcal{B} \rangle$ it follows that $T = I_P(f) \in \langle \mathcal{B} \rangle_u^-$. This establishes that $\langle \mathcal{B} \rangle_u^- = \langle \mathcal{B} \rangle_s^-$. Then Theorem III.2 ensures that $\langle \mathcal{B} \rangle_s^-$ is inverse closed in $\mathcal{L}(X)$. ∎

The following example shows that the inclusion $\langle \mathcal{B} \rangle_u^- \subseteq \langle \mathcal{B} \rangle_s^-$ may be *strict* if $\mathcal{B}$ is only Bade σ-complete.

Example 21. Let $\mathcal{B}$ be the Bade σ-complete B.a. of Example 20(b). It was shown (!) in Exercise 48 that there exists a projection $B \in \langle \mathcal{B} \rangle_w^- = \langle \mathcal{B} \rangle_s^-$ such that $B \notin \mathcal{B}$. By Exercise 31 the projection $B \notin \langle \mathcal{B} \rangle_u^-$ and so the inclusion $\langle \mathcal{B} \rangle_u^- \subseteq \langle \mathcal{B} \rangle_s^-$ is strict. ∎

We make a slight digression. Let X be a Banach space and $m : \Sigma \longrightarrow X$ be a *vector measure*, where Σ is a σ-algebra of subsets of some non-empty set. For each $E \in \Sigma$, let $[E] := \{F \in \Sigma : E \triangle F \text{ is } m\text{-null}\}$. Then the space $\Sigma(m)$ of all such equivalence classes $\{[E] : E \in \Sigma\}$ is an abstractly σ-complete B.a. Considering $\Sigma(m)$ as a subset of $L^1(m)$ we may consider the topology and uniform structure from $L^1(m)$ relativized to $\Sigma(m)$. Then a net $\{[E_\alpha]\} \subseteq \Sigma(m)$ is Cauchy if for every $\varepsilon > 0$ there is $\alpha(\varepsilon)$ such that $\|m\|(E_\alpha \triangle E_\beta) < \varepsilon$ for all $\alpha, \beta \geq \alpha(\varepsilon)$, where $\|m\|(\cdot)$ is the semivariation of m. It is known that $\Sigma(m)$ is always a *complete* uniform space, that is, m is a *closed measure*, [27; Theorem 1, p.78]. Combining this fact with [11; Proposition 1.1] gives the following result.

Lemma V.3. *Let X be a Banach space and $m : \Sigma \longrightarrow X$ be a vector measure. Then the B.a. $\Sigma(m)$ is abstractly complete and, whenever a net $\{[E_\alpha]\} \subseteq \Sigma(m)$ is downwards filtering to $[\emptyset]$ in the order of $\Sigma(m)$, then $\lim_\alpha m(E_\alpha) = 0$ in the norm of X.*

Now, back to B.a. 's of projections.

Definition V.4.   Let X be a Banach space and $\mathcal{B} \subseteq \mathcal{L}(X)$ be a B.a. of projections.

(i) For each $x \in X$ the *cyclic space* generated by x with respect to $\mathcal{B}$ is the closed subspace $\mathcal{B}[x]$, of X, defined to be the closure of the linear span of $\{Bx : B \in \mathcal{B}\}$.

(ii) A vector $x \in X$ is said to be a *cyclic vector* for $\mathcal{B}$ if $X = \mathcal{B}[x]$.   ∎

Example 22.   (a) Let X be any Banach space of dimension at least two and $\mathcal{B} := \{0, I\}$. Given any non-zero vector $x \in X$ it is clear that $\mathcal{B}[x]$ is the 1-dimensional subspace spanned by x. Hence, $\mathcal{B}$ has no cyclic vectors.

(b) Let $X = \mathbb{C}^2$ and $\mathcal{B} := \{0, I, A, B\}$, where $A = \begin{bmatrix} 1 & 0 \\ 0 & 0 \end{bmatrix}$ and $B = \begin{bmatrix} 0 & 0 \\ 0 & 1 \end{bmatrix}$. Then $\mathcal{B}$ is a B.a. of projections with the property that any vector $x = \binom{x_1}{x_2}$ with $x_1 \neq 0$ and $x_2 \neq 0$ is a cyclic vector for $\mathcal{B}$.   ∎

It will be seen later that many proofs of results about B.a. 's of projections reduce the argument first to the case when a cyclic vector is present and then establish the result for that special case. Theorem V.7 below turns out to be useful in this context.

Let $P : \Sigma \longrightarrow \mathcal{L}_s(X)$ be a spectral measure . Fix $x \in X$. Since $P(E \cap F) = P(E)P(F)$ for every $E, F \in \Sigma$, it follows from Proposition I.2 that $P(E)x = 0$ if and only if $\|Px\|(E) = 0$. Hence, the vector measure $Px : \Sigma \longrightarrow X$ has the *special property* that $E \in \Sigma$ is a Px-null set if and only if $(Px)(E) := P(E)x = 0$. For a general vector measure $m : \Sigma \longrightarrow X$ we point out that $m(E) = 0$ alone does *not* imply that E is an m-null set, even in the simplest case when $X = \mathbb{C}$! Indeed, let $\Omega = \{1, 2\}$, let $\Sigma = 2^{\Omega}$ and let $m : \Sigma \longrightarrow \mathbb{C}$ be defined by $m := \delta_2 - \delta_1$, where δ_j is the Dirac point measure at $j \in \Omega$. Then $m(\Omega) = 0$, but there certainly exists a set $F \in \Sigma$ with $F \subseteq \Omega$ such that $m(F) \neq 0$. Hence, Ω is *not* an m-null set.

Theorem V.7.   *Let X be a Banach space and $\mathcal{B} \subseteq \mathcal{L}(X)$ be a Bade σ-complete B.a. of projections with a cyclic vector . Then $\mathcal{B}$ is actually Bade complete .*

Proof.   Let $P : \Sigma \longrightarrow \mathcal{L}_s(X)$ be any spectral measure satisfying $\mathcal{B} = P(\Sigma)$; see Theorem IV.1. Let $x \in X$ be a cyclic vector for $\mathcal{B}$, that is, $X = \mathcal{B}[x] = P(\Sigma)[x]$. Suppose that $E \in \Sigma$ is a Px-null set, that is, $P(E)x = 0$. Then also $P(E)z = 0$ for every z from the linear span of $\{Bx : B \in \mathcal{B}\}$. Since $P(E)$ is continuous and the set of all such vectors z is dense in $X = \mathcal{B}[x]$, it follows that $P(E) = 0$ in $\mathcal{L}(X)$. Hence, the spectral measure P and the vector measure $Px : \Sigma \longrightarrow X$ have the same null sets. Accordingly, $\Sigma(P)$ and $\Sigma(Px)$ are isomorphic as B.a. 's. Since $\Sigma(Px)$ is abstractly complete (c.f. Lemma V.3), so is $\Sigma(P)$.

Now, let $\{[E_\alpha]\} \subseteq \Sigma(P)$ be downwards filtering to $[\emptyset]$, in which case $\{[E_\alpha]\}$ is also downwards filtering to $[\emptyset]$ in $\Sigma(Px)$. By Lemma V.3 we have that $P(E_\alpha)x = Px(E_\alpha) \longrightarrow 0$ in the norm of X. Then also $P(E_\alpha)z \longrightarrow 0$ in X for every z in the linear span of $\{Bx : B \in \mathcal{B}\}$. Since $\sup_\alpha \|P(E_\alpha)\| \leq \|\mathcal{B}\| < \infty$, it follows from the density of such vectors z that $P(E_\alpha)u \longrightarrow 0$ for all $u \in X$, that is, $P(E_\alpha) \longrightarrow 0$ in $\mathcal{L}_s(X)$. Then Lemma IV.3 implies that P is a *closed* spectral measure and hence, $\mathcal{B} = P(\Sigma)$ is Bade complete ; see Theorem IV.1.   ∎

The following technical result will prove to be quite useful.

Lemma V.4.   *Let X be a Banach space and $\mathcal{B} \subseteq \mathcal{L}(X)$ be a Bade σ-complete B.a. of* *projections. Suppose that Y is a closed subspace of X such that $BY \subseteq Y$ for every $B \in \mathcal{B}$. Let $\mathcal{B}_Y \subseteq \mathcal{L}(Y)$ denote the collection of all restrictions $B_Y : Y \longrightarrow Y$ of elements $B \in \mathcal{B}$. Then $\mathcal{B}_Y$ is a Bade σ-complete B.a. of projections in $\mathcal{L}(Y)$.*

Proof. By Theorem IV.1 there exists a spectral measure $P : \Sigma \longrightarrow \mathcal{L}_s(X)$ such that $\mathcal{B} = P(\Sigma)$. If $y \in Y \subseteq X$, then for each $E \in \Sigma$ we have that $(P(E)|_Y)y = P(E)y$. Hence, if $\{E_n\}_{n=1}^\infty \subseteq \Sigma$ and $E_n \downarrow \emptyset$, then $(P(E_n)|_Y)y \longrightarrow 0$ in Y. This shows that $P_Y : \Sigma \longrightarrow \mathcal{L}_s(Y)$ given by $E \mapsto P(E)|_Y$ is σ-additive. Since it is routine to verify that $P_Y(\emptyset) = 0$, $P_Y(\Omega) = I_Y$ and $P_Y(E \cap F) = P_Y(E)P_Y(F)$ for each $E, F \in \Sigma$, it follows that P_Y is a spectral measure. So, by Theorem IV.1 again we deduce that $\mathcal{B}_Y = P_Y(\Sigma)$ is a Bade σ-complete B.a. ∎

Exercise 49. Let X be a Banach space and $\{P_\alpha\}_{\alpha \in A} \subseteq \mathcal{L}(X)$ be a net of *commuting* projections. Suppose that $\sup\{\|P_\alpha\| : \alpha \in A\} < \infty$ and that $P = \lim_\alpha P_\alpha$ exists in $\mathcal{L}_s(X)$. Show that P is a projection and $PP_\alpha = P_\alpha P$ for all $\alpha \in A$. ∎

Exercise 50. Let X be a Banach space and $\mathcal{B} \subseteq \mathcal{L}(X)$ be a *bounded* B.a. of projections. Let $\overline{\mathcal{B}}$ denote the closure of $\mathcal{B}$ in $\mathcal{L}_s(X)$.

(a) Show that $\overline{\mathcal{B}}$ is a bounded subset of $\mathcal{L}_u(X)$.

(b) Show that $\overline{\mathcal{B}}$ is again a B.a. of projections. ∎

Given a (general) B.a. $\mathcal{B}$ which is abstractly σ-complete , it is not always easy to identify its abstract completion in a concrete manner in terms of $\mathcal{B}$ itself. However, if $\mathcal{B}$ is a Bade σ-complete B.a. of *projections*, then the following result identifies the Bade completion of $\mathcal{B}$ in a very direct way in terms of $\mathcal{B}$ itself.

Theorem V.8. *Let X be a Banach space and $\mathcal{B} \subseteq \mathcal{L}(X)$ be a Bade σ-complete B.a. of projections. Let $\overline{\mathcal{B}}$ denote the topological closure of $\mathcal{B}$ in $\mathcal{L}_s(X)$. Then $\overline{\mathcal{B}}$ is a Bade complete B.a. of projections .*

Proof. Since $\mathcal{B}$ is bounded (c.f. Theorem III.1) so is $\overline{\mathcal{B}}$; see Exercise 50(a). By Exercise 50(b) $\overline{\mathcal{B}}$ is again a B.a. of projections. Suppose that $\overline{\mathcal{B}}$ is *not* Bade complete. By Theorem IV.1 and Lemma IV.1 there is a monotone increasing net $\{\overline{B}_\alpha\} \subseteq \overline{\mathcal{B}}$ and $x \in X$ such that $\{\overline{B}_\alpha x\}$ is not a convergent net in X. Let $\mathcal{B}[x]$ and $\overline{\mathcal{B}}[x]$ be the cyclic spaces generated by x with respect to $\mathcal{B}$ and $\overline{\mathcal{B}}$, respectively. If $\overline{B} \in \overline{\mathcal{B}}$, then $\overline{B} = \lim A_\beta$ in $\mathcal{L}_s(X)$ for some net $\{A_\beta\} \subseteq \mathcal{B}$ and so $\overline{B}x = \lim_\beta A_\beta x$ belongs to $\mathcal{B}[x]$, because each vector $A_\beta x \in \mathcal{B}[x]$ and $\mathcal{B}[x]$ is closed in X. Accordingly, $\{\overline{B}x : \overline{B} \in \overline{\mathcal{B}}\} \subseteq \mathcal{B}[x]$ and it follows that $\overline{\mathcal{B}}[x] \subseteq \mathcal{B}[x]$. Since $\mathcal{B} \subseteq \overline{\mathcal{B}}$, it is also clear that $\mathcal{B}[x] \subseteq \overline{\mathcal{B}}[x]$ and hence $\mathcal{B}[x] = \overline{\mathcal{B}}[x]$. So, if $Y := \mathcal{B}[x]$, then it is routine to check that $\overline{B}Y = \overline{B}(\overline{\mathcal{B}}[x]) \subseteq \overline{\mathcal{B}}[x] = Y$, for each $\overline{B} \in \overline{\mathcal{B}}$. For each $\overline{B} \in \overline{\mathcal{B}}$, let $\overline{B}_Y \in \mathcal{L}(Y)$ denote the restriction of $\overline{B}$ to Y. It can be verified that $\overline{\mathcal{B}}_Y := \{\overline{B}_Y : \overline{B} \in \overline{\mathcal{B}}\}$ is contained in the closure of $\{B_Y : B \in \mathcal{B}\}$ in $\mathcal{L}_s(Y)$; see Exercise 51 below.

Since $\{\overline{B}_\alpha x\}$ is not convergent in X, it follows that $\{\overline{B}_\alpha|_Y x\}$ is not convergent in Y since $x \in Y$ and $\overline{B}_\alpha x = (\overline{B}_\alpha|_Y)x$, for all α. That is, $\{\overline{B}_\alpha|_Y\}$ is not convergent in $\mathcal{L}_s(Y)$ and hence, by the previous paragraph, the closure of $\{B_Y : B \in \mathcal{B}\}$ in $\mathcal{L}_s(Y)$ contains the monotone increasing net $\{\overline{B}_\alpha|_Y\}$ which fails to have a limit in $\mathcal{L}_s(Y)$. By Theorem IV.1 the closure of $\{B_Y : B \in \mathcal{B}\}$ in $\mathcal{L}_s(Y)$ cannot be Bade complete . But, by Lemma V.4 it is Bade σ-complete and so, by Theorem V.7 applied in $Y = \mathcal{B}[x]$, it is also Bade complete. This contradiction shows the initial assumption that $\overline{\mathcal{B}}$ is not Bade complete is false. Hence, $\overline{\mathcal{B}}$ is Bade complete .   ∎

Exercise 51.   In the notation of the proof of Theorem V.8 show that $\mathcal{B}_Y$ is contained in the closure of $\{B_Y : B \in \mathcal{B}\}$ in $\mathcal{L}_s(Y)$.   ∎

As an immediate consequence of the previous theorem we have the following fact.

Corollary V.8.1.   *Let X be a Banach space and $\mathcal{B} \subseteq \mathcal{L}(X)$ be a Bade σ-complete B.a. of projections. Then $\mathcal{B}$ is Bade complete if and only if $\mathcal{B}$ is a closed subset of $\mathcal{L}_s(X)$.*

Proof.   If $\mathcal{B}$ is Bade complete, then it is a closed subset of $\mathcal{L}_s(X)$ by Corollary V.5.1. On the other hand, if $\mathcal{B}$ is closed in $\mathcal{L}_s(X)$, then $\overline{\mathcal{B}} = \mathcal{B}$ and so $\mathcal{B}$ is Bade complete by Theorem V.8.   ∎

Remark.   In Exercise 47 it is shown that there exists an abstractly complete B.a. of projections $\mathcal{B} \subseteq \mathcal{L}(\ell^\infty)$ such that $\mathcal{B}$ is closed in $\mathcal{L}_s(X)$, but $\mathcal{B}$ fails to be Bade complete . Hence, the assumption that $\mathcal{B}$ is Bade σ-complete in Corollary V.8.1 *cannot* be replaced by the requirement that $\mathcal{B}$ is abstractly complete. The interested reader may wish to look at the paper [22] where it is shown that if X does not contain an isomorphic copy of c_0, then every bounded B.a. of projections $\mathcal{B} \subseteq \mathcal{L}(X)$ which is a closed subset of $\mathcal{L}_s(X)$ is necessarily Bade complete.   ∎

Exercise 52.   Let X be a Banach space and $\mathcal{B} \subseteq \mathcal{L}(X)$ be a Bade σ-complete B.a. of projections.

(a) Show that if $A \in \mathcal{L}(X)$ is a projection such that $A \in \langle \mathcal{B} \rangle_w^-$, then $A \in \overline{\mathcal{B}}$ (the closure of $\mathcal{B}$ in $\mathcal{L}_s(X)$).

(b) Show that $\langle \mathcal{B} \rangle_u^- \subseteq \langle \mathcal{B} \rangle_s^- = \langle \mathcal{B} \rangle_w^-$ is valid (compare with Corollary V.6.3).

(c) Show that $\langle \mathcal{B} \rangle_s^- = \langle \overline{\mathcal{B}} \rangle_s^-$, where again $\overline{\mathcal{B}}$ is the closure of $\mathcal{B}$ in $\mathcal{L}_s(X)$.   ∎

Recall for a Hilbert space H that an operator $T \in \mathcal{L}(H)$ is called *normal* if $TT^* = T^*T$, where T^* denotes the (Hilbert space) adjoint operator. If $T = T^*$, then T is called *selfadjoint*

Proposition V.5.   *Let H be a Hilbert space.*

(a) *Let $T \in \mathcal{L}(H)$ be a normal operator. Then $AT^* = T^*A$ whenever $A \in \mathcal{L}(H)$ satisfies $AT = TA$.*

(b) *Let $\mathcal{B} \subseteq \mathcal{L}(H)$ be a bounded B.a. of projections. Then there exists a selfadjoint operator $S \in \mathcal{L}(H)$ which is invertible in $\mathcal{L}(H)$, such that the B.a. of projections $\{SBS^{-1} : B \in \mathcal{B}\}$ consists entirely of selfadjoint projections.*

Part (a) is the *Fuglede theorem*, [13; Theorem 7.21], and part (b) is the well known *Mackey-Werner theorem* , [13; Proposition 8.2].

The following example shows that if $\mathcal{B} \subseteq \mathcal{L}(H)$ is a B.a. of selfadjoint projections, then the elements of $\langle \mathcal{B} \rangle_s^-$ are rather special.

Example 23.   Let H be a Hilbert space and $\mathcal{B} \subseteq \mathcal{L}(H)$ be a B.a. of *selfadjoint* projections, in which case $\|\mathcal{B}\| = 1$, [13; Proposition 7.14]. The claim is that $\langle \mathcal{B} \rangle_s^-$ consists entirely of *normal* operators. To see this observe that every operator of the form $\sum_{j=1}^n \alpha_j B_j$ with $\alpha_j \in \mathbb{C}$ and $B_j \in \mathcal{B}$, for $1 \leq j \leq n$, is clearly normal. Since every operator in $\langle \mathcal{B} \rangle_s^-$ is the limit in $\mathcal{L}_s(H)$ of a net of such operators it suffices to establish the following

Fact. *Let $\{T_\alpha\} \subseteq \mathcal{L}(H)$ be a net of pairwise commuting, normal operators such that $T = \lim_\alpha T_\alpha$ exists in $\mathcal{L}_s(H)$. Then T is normal.*

To establish this Fact fix an index α. It is routine to check that $TT_\alpha = T_\alpha T$ and so, by Fuglede's theorem above, $TT_\alpha^* = T_\alpha^* T$. Since also $T_\alpha \longrightarrow T$ in $\mathcal{L}_w(H)$- see Exercise 32(a)- it follows for each $x, y \in H$ that

$$(17) \qquad \lim_\alpha \langle T_\alpha x, Ty \rangle = \langle Tx, Ty \rangle = \langle T^*Tx, y \rangle.$$

But, again by Fuglede's theorem,

$$(18) \qquad \langle T_\alpha x, Ty \rangle = \langle x, T_\alpha^* Ty \rangle = \langle x, TT_\alpha^* y \rangle = \langle T^* x, T_\alpha^* y \rangle = \langle T_\alpha T^* x, y \rangle.$$

Since $T_\alpha \longrightarrow T$ in $\mathcal{L}_w(H)$, we have from (18) that $\langle T_\alpha x, Ty \rangle = \langle T_\alpha T^* x, y \rangle \xrightarrow{\alpha} \langle TT^* x, y \rangle$. It then follows from (17) that $\langle T^*Tx, y \rangle = \langle TT^* x, y \rangle$, for all $x, y \in H$, and hence $T^*T = TT^*$. That is, T is normal . ∎

Definition V.5. Let X be a Banach space. An operator $T \in \mathcal{L}(X)$ is called a *scalar-type spectral operator* if there exists some spectral measure $P : \Sigma \longrightarrow \mathcal{L}_s(H)$ and a P-integrable function f such that $T = \int_\Omega f dP$. ∎

Given an operator $T \in \mathcal{L}(X)$ recall that the *spectrum* $\sigma(T)$, of T, consists of all numbers $\lambda \in \mathbb{C}$ such that $(T - \lambda I)$ is *not* invertible in $\mathcal{L}(X)$. In the notation of Definition V.5 if $T \in \mathcal{L}(X)$ is a scalar-type spectral operator, then it turns out that

$$(19) \qquad \sigma(T) = \cap\{\overline{f(E)} : E \in \Sigma, \ P(E) = I\},$$

where $\overline{f(E)}$ is the closure (in $\mathbb{C}$) of $f(E) := \{f(w) : w \in E\}$. Indeed, if $E \in \Sigma$ and $P(E) = I$, then we define, for each $\lambda \notin \overline{f(E)}$, the function

$$h_\lambda(w) := (\lambda - f(w))^{-1} \chi_E(w), \qquad w \in \Omega.$$

Observe that $h_\lambda \in L^\infty(P)$ and satisfies

$$\left(\int_\Omega h_\lambda \, dP\right)(\lambda I - T) = \left(\int_\Omega h_\lambda \, dP\right)\left(\int_\Omega (\lambda - f) \, dP\right) = \int_\Omega \chi_E \, dP = P(E) = I.$$

Accordingly, $(\lambda I - T)^{-1} = \int_\Omega h_\lambda \, dP$ exists in $\mathcal{L}(X)$ and so λ belongs to the *resolvent set* $\rho(T) := \mathbb{C}\backslash\sigma(T)$, of T. That is, $\sigma(T) \subseteq \overline{f(E)}$ whenever $P(E) = I$ and so

$$\cap\{\overline{f(E)} : E \in \Sigma, \ P(E) = I\} \supseteq \sigma(T).$$

Conversely, if $\lambda \in \rho(T) = \rho(\int_\Omega f dP)$, then $\int_\Omega (\lambda - f) dP = (\lambda I - T)$ is invertible in $\mathcal{L}(X)$ with $(\lambda - f) \in L^1(P) = L^\infty(P)$ (as vector spaces) . By Theorem V.1 the function $(\lambda - f)^{-1} : w \mapsto 1/(\lambda - f(w))$, for $w \in \Omega$, is an element of $L^\infty(P)$. Hence, there is a set $E_0 \in \Sigma$ with $P(E_0) = I$ such that $(1/|\lambda - f(w)|) \leq M$, for $w \in E_0$, for some $M > 0$. Then $\lambda \notin \overline{f(E_0)}$ and so $\lambda \notin \cap\{\overline{f(E)} : E \in \Sigma, \ P(E) = I\}$, which shows that $\sigma(T) \supseteq \cap\{\overline{f(E)} : E \in \Sigma, \ P(E) = I\}$.

Hence, (19) is indeed valid and is, moreover, independent of P and f in the sense that P and f only need to satisfy $T = \int_\Omega f dP$.

Define $P_T : Bo(\mathbb{C}) \longrightarrow \mathcal{L}_s(X)$ by $P_T(\delta) = P(f^{-1}(\delta))$, for each $\delta \in Bo(\mathbb{C})$. It is routine to verify that P_T is a spectral measure and satisfies $P_T(\delta) = 0$ if $\delta \subseteq \rho(T)$; see (19). Moreover, the identity function on $\mathbb{C}$ is P_T-essentially bounded (as $\sigma(T)$ is compact and $P_T(\sigma(T)) = I$) and satisfies

$$T = \int_{\mathbb{C}} \lambda \, dP_T(\lambda) = \int_{\sigma(T)} \lambda \, dP_T(\lambda).$$

The spectral measure P_T is called the *resolution of the identity* of T.

The following result shows that in Hilbert spaces the scalar-type spectral operators are essentially the normal operators.

Proposition V.6. (a) *Let H be a Hilbert space and $T \in \mathcal{L}(H)$ be a scalar-type spectral operator. Then there exists a selfadjoint operator $S \in \mathcal{L}(H)$ which is invertible in $\mathcal{L}(H)$, such that STS^{-1} is a normal operator.*

(b) *Let $\mathcal{B} \subseteq \mathcal{L}(H)$ be a Bade σ-complete B.a. of projections. Then every element of $\langle \mathcal{B} \rangle_s^-$ is a scalar-type spectral operator.*

Proof. (a) Let $P : \Sigma \longrightarrow \mathcal{L}_s(H)$ be a spectral measure and $f \in L^1(P)$ satisfy $T = \int_\Omega f dP$. Then $P(\Sigma)$ is a bounded B.a. of projections (see Lemma III.3) and so, by the Mackey-Wermer theorem (c.f. Proposition V.5(b)), there exists an invertible, selfadjoint operator $S \in \mathcal{L}(H)$ such that $\{SP(E)S^{-1} : E \in \Sigma\}$ consists entirely of selfadjoint projections. Then $R : \Sigma \longrightarrow \mathcal{L}_s(H)$ defined by $R(E) = SP(E)S^{-1}$, for each $E \in \Sigma$, is a selfadjoint-valued spectral measure satisfying

$$STS^{-1} = S(\int_\Omega f dP)S^{-1} = \int_\Omega f dR \in \langle R(\Sigma) \rangle_s^-.$$

By putting $\mathcal{B} := R(\Sigma)$ in Example 23 it follows that STS^{-1} is normal.

(b) By Theorem V.8 the closure $\overline{\mathcal{B}}$, of $\mathcal{B}$, in $\mathcal{L}_s(H)$ is a Bade complete B.a. of projections. By Theorem IV.1 there exists some *closed* spectral measure $P : \Sigma \longrightarrow \mathcal{L}_s(H)$ such that $\overline{\mathcal{B}} = P(\Sigma)$. Let $T \in \langle \mathcal{B} \rangle_s^-$. Then also $T \in \langle \overline{\mathcal{B}} \rangle_s^-$; c.f. Exercise 52(c). So, Theorem V.6 implies that $T = \int_\Omega f dP$ for some $f \in L^1(P)$, i.e. T is a scalar-type spectral operator. ∎

An examination of the proof of Proposition V.6(b) shows that it is irrelevant that the underlying space is a Hilbert space. Hence, the same proof establishes the final result of this chapter.

Proposition V.7. *Let X be a Banach space and $\mathcal{B} \subseteq \mathcal{L}(X)$ be a Bade σ-complete B.a. of projections. Then every element of $\langle \mathcal{B} \rangle_w^- = \langle \mathcal{B} \rangle_s^-$ is a scalar-type spectral operator.*

Recall that an operator $S \in \mathcal{L}(X)$ is called *quasinilpotent* if $\sigma(S) = \{0\}$ or, equivalently, if $\lim_{n \to \infty} \|S^n\|^{1/n} = 0$. Let $T \in \mathcal{L}(X)$ be a scalar-type spectral operator, say $T = \int_\Omega f dP$ in the notation of Definition V.5. It is clear from (19) that if T is also quasinilpotent, then $f = 0$, P-a.e. , and so $T = 0$. If then follows from Proposition V.7 that the closed subalgebra $\langle \mathcal{B} \rangle_s^- = \langle \mathcal{B} \rangle_w^-$ of $\mathcal{L}(X)$ is *semisimple* (i.e. contains no non-zero quasinilpotent elements) whenever $\mathcal{B} \subseteq \mathcal{L}(X)$ is a Bade σ-complete B.a. of projections.

Chapter VI

Bade functionals: an application to scalar-type spectral operators

Let H be a Hilbert space and $\mathcal{B} \subseteq \mathcal{L}(H)$ be a B.a. of *selfadjoint* projections. Fix $x \in H$. Using the fact that $B^2 = B = B^*$ and $\|Bx\|^2 = \langle Bx, Bx \rangle$, for each $B \in \mathcal{B}$, the calculation

$$\langle Bx, x \rangle = \langle B^2 x, x \rangle = \langle Bx, B^* x \rangle = \langle Bx, Bx \rangle = \|Bx\|^2$$

shows that $\langle Bx, x \rangle \geq 0$, for $B \in \mathcal{B}$, and that $Bx = 0$ whenever $B \in \mathcal{B}$ satisfies $\langle Bx, x \rangle = 0$.

Suppose now that $\mathcal{B} \subseteq \mathcal{L}(H)$ is an arbitrary bounded B.a. of projections. Let $S \in \mathcal{L}(H)$ be an invertible , selfadjoint operator such that $\mathcal{A} := \{SBS^{-1} : B \in \mathcal{B}\}$ is a B.a. of selfadjoint projections; see Proposition V.5(b). Fix $x \in H$ and let $x' := S^2 x$. For each $B \in \mathcal{B}$ let $B^\# := SBS^{-1} \in \mathcal{A}$, in which case $(B^\#)^2 = B^\# = (B^\#)^*$. The calculation

$$
\begin{aligned}
\langle Bx, x' \rangle &= \langle Bx, S^2 x \rangle = \langle S^* Bx, Sx \rangle = \langle SBx, Sx \rangle = \langle B^\# Sx, Sx \rangle \\
&= \langle (B^\#)^2 Sx, Sx \rangle = \langle B^\# Sx, (B^\#)^* Sx \rangle = \langle B^\# Sx, B^\# Sx \rangle = \|B^\# Sx\|^2,
\end{aligned}
$$

valid for each $B \in \mathcal{B}$, shows that $\langle Bx, x' \rangle \geq 0$, for all $B \in \mathcal{B}$. Moreover, if $B \in \mathcal{B}$ satisfies $\langle Bx, x' \rangle = 0$, then the previous calculation shows that $0 = B^\# Sx = SBx$. Since S is injective it follows that $Bx = 0$. So, identifying H' with H and interpreting $x' = S^2 x \in H'$ we have shown, for each $x \in H$, that there exists $x' \in H'$ with the properties $\langle Bx, x' \rangle \geq 0$, for all $B \in \mathcal{B}$, and $Bx = 0$ whenever $B \in \mathcal{B}$ satisfies $\langle Bx, x' \rangle = 0$.

W. Bade showed in [1] that, remarkably, this property of bounded B.a. 's of projections in a Hilbert space carries over to Bade σ-complete B.a. 's of projections in Banach spaces. One of the aims of this chapter is to establish Bade's result. Our proof, based on Rybakov's theorem for vector measures, is quite different to Bade's original proof. For still another proof see [23]. Of course, Rybakov's theorem appeared some 15 years after Bade's proof! In the second part of the chapter we give a non-trivial application of Bade functionals to show that every Bade σ-complete B.a. of projections in a *separable* Banach space coincides with the resolution of the identity of some scalar-type spectral operator with real spectrum , [31].

Definition VI.1. Let X be a Banach space and $\mathcal{B} \subseteq \mathcal{L}(X)$ be a bounded B.a. of projections. Given $x \in X$, any non-zero vector $x' \in X'$ (if it exists) with the properties

(i) $\langle Bx, x' \rangle \geq 0$, for all $B \in \mathcal{B}$,

and

(ii) $Bx = 0$ whenever $B \in \mathcal{B}$ satisfies $\langle Bx, x' \rangle = 0$,

is called a *Bade functional* for x with respect to $\mathcal{B}$.   ∎

The existence of Bade functionals is guaranteed by the next result.

Theorem VI.1.   *Let X be a Banach space and $\mathcal{B} \subseteq \mathcal{L}(X)$ be any Bade σ-complete B.a. of projections. Then every $x \in X$ has a Bade functional with respect to $\mathcal{B}$.*

Proof.   Fix $x \in X$. Let $Y := \mathcal{B}[x]$ be the cyclic space generated by x with respect to $\mathcal{B}$. By Lemma V.4 the restricted B.a. $\mathcal{B}_Y \subseteq \mathcal{L}(Y)$ of $\mathcal{B}$ to Y is also a Bade σ-complete B.a. of projections. Since $x \in Y$ is a cyclic vector for $\mathcal{B}_Y$, it follows from Theorem V.7 and Theorem IV.1 that there exists a *closed* spectral measure $P : \Sigma \longrightarrow \mathcal{L}_s(Y)$ satisfying $P(\Sigma) = \mathcal{B}_Y$. Then Rybakov's theorem (see Theorem I.10) applied to the vector measure $Px : \Sigma \longrightarrow Y$ guarantees the existence of a unit vector $y' \in Y'$ such that $Px \ll |\langle Px, y' \rangle|$. Since $\langle Px, y' \rangle \ll |\langle Px, y' \rangle|$, the Radon-Nikodym theorem for scalar measures (c.f. Theorem I.6) guarantees the existence of a Σ-measurable function $\phi : \Omega \longrightarrow \mathbb{C}$ satisfying $|\phi(w)| = 1$, for $|\langle Px, y' \rangle|$-a.e. $w \in \Omega$, such that

$$(1) \qquad\qquad |\langle Px, y' \rangle|(\delta) = \int_\delta \phi \, d\langle Px, y' \rangle, \qquad \delta \in \Sigma.$$

Define ϕ to be zero on the $|\langle Px, y' \rangle|$-null set E for which $|\phi| \neq 1$. Since E is then also Px-null and x is a cyclic vector for $P(\Sigma) = \mathcal{B}_Y$, it follows that E is also P-null . Accordingly, $\phi \in L^\infty(P)$. Define $T := \int_\Omega \phi \, dP \in \mathcal{L}(Y)$. Finally, let $x' \in X'$ be any continuous linear functional on X which coincides with $T'y' \in Y'$ on the closed subspace $Y \subseteq X$; see Theorem I.5(f).

Fix $B \in \mathcal{B}$. Then $B|_Y = P(\delta)$ for some $\delta \in \Sigma$. Since $x \in Y$, we have

$$\langle Bx, x' \rangle = \langle P(\delta)x, x' \rangle = \langle P(\delta)x, T'y' \rangle$$

where that last equality uses the fact that $P(\delta)x \in Y$ and $x'|_Y \equiv T'y'$. But,

$$\begin{aligned}
\langle P(\delta)x, T'y' \rangle &= \langle P(\delta)x, (\int_\Omega \phi \, dP)'y' \rangle = \langle P(\delta)(\int_\Omega \phi \, dP)x, y' \rangle \\
&= \langle (\int_\delta \phi \, dP)x, y' \rangle = \int_\delta \phi \, d\langle Px, y' \rangle = |\langle Px, y' \rangle|(\delta),
\end{aligned}$$

where the last equality follows from (1). Hence, $\langle Bx, x' \rangle = |\langle Px, y' \rangle|(\delta) \geq 0$ and so (i) of Definition VI.1 is satisfied.

Suppose that $B \in \mathcal{B}$ satisfies $\langle Bx, x' \rangle = 0$. Then also $\langle B_Y x, T'y' \rangle = 0$, as $Bx \in Y$ and $x'|_Y \equiv T'y'$. Since $B_Y = P(E_0)$ for some $E_0 \in \Sigma$ the same calculation as above shows that

$$0 = \langle Bx, x' \rangle = |\langle Px, y' \rangle|(E_0).$$

Since $Px \ll |\langle Px, y' \rangle|$, we deduce that $Bx = P(E_0)x = 0$. Accordingly, (ii) of Definition VI.1 is also satisfied.   ∎

We point out that Bade σ-complete B.a.'s of projections are not the only ones having a Bade functional for each $x \in X$.

Example 24. Let $X := \ell^\infty$ and $\mathcal{B} \subseteq \mathcal{L}(X)$ be the B.a. consisting of all projections $P(E) \in \mathcal{L}(X)$ given by

$$P(E) : x \mapsto (\chi_E(1)x_1, \chi_E(2)x_2, \dots), \qquad x = (x_1, x_2, \dots) \in X,$$

where E is an arbitrary subset of $\mathbb{N}$. Then $\mathcal{B}$ is abstractly complete but not Bade σ-complete . Given any non-zero vector $x \in X$, let x' be the sequence defined by $(x')_n := \bar{x}_n / 2^n \|x\|_\infty$, for each $n \in \mathbb{N}$. Then $x' \in \ell^1 \subseteq X'$ and $\langle P(E)x, x' \rangle = \sum_{n=1}^\infty (|x_n|^2 / 2^n \|x\|_\infty) \chi_E(n)$ for each set $E \subseteq \mathbb{N}$. It is then clear from Definition VI.1 that x' is a Bade functional for x with respect to $\mathcal{B}$. ∎

The existence of Bade functionals in Example 24 is due to the countability of the set $\mathbb{N}$ used to form the Banach space $\ell^\infty = \ell^\infty(\mathbb{N})$. Removing this condition provides an example of an abstractly complete B.a. of projections for which *not* every vector has a Bade functional.

Example 25. Let Λ be any uncountable set and Σ be the family of all subsets of Λ. Let $X := \ell^\infty(\Lambda)$ be the Banach space of all bounded functions $f : \Lambda \longrightarrow \mathbb{C}$ equipped with the norm $\|f\|_\infty := \sup\{|f(\lambda)| : \lambda \in \Lambda\}$. Then the dual Banach space X' consists of all finitely additive set functions $\mu : \Sigma \longrightarrow \mathbb{C}$ for which $\sup\{|\mu(E)| : E \in \Sigma\} < \infty$; see [14; Ch.IV, §5]. Let $\mathcal{B}$ be the B.a. consisting of all projections $P(E) \in \mathcal{L}(X)$ defined by $P(E)f = \chi_E f$, for $f \in X$ and each $E \in \Sigma$. It is routine to verify that $P : \Sigma \longrightarrow \mathcal{B}$ is a B.a. isomorphism. Since Σ is abstractly complete the B.a. $\mathcal{B}$ is also abstractly complete .

Consider the particular element $\mathbb{1} \in X$ (i.e. the function constantly equal to 1 on Λ). Suppose that $\mu \in X'$ is a Bade functional for $\mathbb{1}$ with respect to $\mathcal{B}$, in which case

$$0 \leq \langle P(E)\mathbb{1}, \mu \rangle = \int_E \mathbb{1} \, d\mu = \mu(E), \qquad E \in \Sigma;$$

see Definition VI.1(i). By Definition VI.1(ii) we see that $\langle P(E)\mathbb{1}, \mu \rangle = 0$ implies $P(E)\mathbb{1} = \chi_E = 0$ in X (i.e. $E = \emptyset$), from which it follows that $\mu(\{\lambda\}) > 0$ for each $\lambda \in \Lambda$. That is, $\{\lambda \in \Lambda : \mu(\{\lambda\}) > 0\} = \Lambda$. But, the finite additivity of μ together with the property $\sup\{|\mu(E)| : E \in \Sigma\} < \infty$ implies that $E_n := \{\lambda \in \Lambda : \mu(\{\lambda\}) \geq \frac{1}{n}\}$ is a finite set, for each $n \in \mathbb{N}$. Accordingly, the set $\cup_{n=1}^\infty E_n = \{\lambda \in \Lambda : \mu(\{\lambda\}) > 0\}$ is countable. But, we saw above that this set equals Λ and we have a contradiction. So, $\mathbb{1}$ cannot have a Bade functional with respect to $\mathcal{B}$. It is interesting to note that $\mathbb{1}$ is a cyclic vector for $\mathcal{B}$. ∎

Given a B.a. of projections $\mathcal{B} \subseteq \mathcal{L}(X)$ and a *particular* vector $x \in X$, it is also of interest to be able to determine when x admits a Bade functional with respect to $\mathcal{B}$. For criteria in relation to this question we refer to [23].

As an immediate application of Theorem VI.1 we present an interesting result where weak operator convergence implies strong operator convergence; that this is not the case in general was noted in Exercise 32(b).

Proposition VI.1. *Let X be a Banach space and $\mathcal{B} \subseteq \mathcal{L}(X)$ be a Bade σ-complete B.a. of projections. Let $\{B_\alpha\}_{\alpha \in A} \subseteq \mathcal{B}$ be a net and $B \in \mathcal{L}(X)$ be a projection such that $\lim_\alpha B_\alpha = B$ in $\mathcal{L}_w(X)$. Then $\lim_\alpha B_\alpha = B$ in $\mathcal{L}_s(X)$.*

Proof. Let $\overline{B}$ be the closure of B in $\mathcal{L}_s(X)$, in which case $\overline{B}$ is a Bade complete B.a.; see Theorem V.8. By Corollary V.6.1 the limit projection $B \in \overline{B}$. Observe that

$$\langle BB_\alpha x, x'\rangle = \langle B_\alpha x, B'x'\rangle \xrightarrow{\alpha} \langle Bx, B'x'\rangle = \langle B^2 x, x'\rangle = \langle Bx, x'\rangle, \qquad x \in X, \ x' \in X',$$

showing that $B_\alpha B \xrightarrow{\alpha} B$ in $\mathcal{L}_w(X)$. Since $B_\alpha B \leq B$ we have, by Exercise 30(b), that $(B - B_\alpha B) \in \mathcal{B}$ for all α, and we just showed that $(B - B_\alpha B) \xrightarrow{\alpha} 0$ in $\mathcal{L}_w(X)$. A similar calculation shows that $(I - B)B_\alpha \xrightarrow{\alpha} 0$ in $\mathcal{L}_w(X)$, since

$$\langle (I - B)B_\alpha x, x'\rangle = \langle B_\alpha x, (I - B)'x'\rangle \xrightarrow{\alpha} \langle Bx, (I - B)'x'\rangle = \langle (I - B)Bx, x'\rangle = 0,$$

for all $x \in X$ and $x' \in X'$. Moreover, for each α the identities

$$B - B_\alpha = (B - B_\alpha)[B + (I - B)] = (B - B_\alpha)B + (B - B_\alpha)(I - B),$$

together with $B(I - B) = 0$ show that

$$B - B_\alpha = (B - B_\alpha B) - (I - B)B_\alpha, \qquad \alpha \in A,$$

is the difference of the nets $\{B - B_\alpha B\}$ and $\{(I - B)B_\alpha\}$, both from $\overline{B}$, which converge to 0 in $\mathcal{L}_w(X)$. So, it suffices to show that if a net $\{D_\alpha\} \subseteq \overline{B}$ satisfies $D_\alpha \xrightarrow{\alpha} 0$ in $\mathcal{L}_w(X)$, then also $D_\alpha \xrightarrow{\alpha} 0$ in $\mathcal{L}_s(X)$. Suppose, by contradiction, that $D_\alpha \not\xrightarrow{} 0$ in $\mathcal{L}_s(X)$. Then there exists $x_0 \in X$ such that $\{D_\alpha x_0\}$ does *not* converge to 0 in X. Let $P : \Sigma \longrightarrow \mathcal{L}_s(X)$ be a *closed* spectral measure such that $P(\Sigma) = \overline{B}$. Then there exist sets $E_\alpha \in \Sigma$ with $P(E_\alpha) = D_\alpha$, for each α. Let $x' \in X'$ be a Bade functional for x_0 with respect to $\overline{B}$. Then $Px_0 \ll \langle Px_0, x'\rangle$. Since $\lim_\alpha \langle P(E_\alpha)x_0, x'\rangle = 0$ it follows from Theorem I.7 (and the discussion prior to that result) that also $\lim_\alpha P(E_\alpha)x_0 = 0$ (in the norm of X) which is contrary to our assumption. Hence, no such x_0 exists. That is, $P(E_\alpha) = D_\alpha \xrightarrow{\alpha} 0$ in $\mathcal{L}_s(X)$.   ∎

We now proceed to give a more involved application.

It is well known that every Bade σ-complete B.a. of *selfadjoint* projections in a *separable Hilbert space* coincides with the resolution of the identity of some selfadjoint operator; see [10; p.134], for example.

Exercise 53. Let H be a *separable* Hilbert space. It was just noted that every Bade σ-complete B.a. of *selfadjoint* projections in H coincides with the resolution of the identity of some selfadjoint operator. Deduce that *any* Bade σ-complete B.a. (of not necessarily selfadjoint) projections in H coincides with the resolution of the identity of some scalar-type spectral operator in H with real spectrum .   ∎

The aim for the remainder of this chapter is to establish the same result in Banach spaces, that is, to show that any Bade σ-complete B.a. of projections in a separable Banach space coincides with the resolution of the identity of some scalar-type spectral operator with real spectrum ; see [31].

We will require a number of preliminary results.

Definition VI.2. A (general) Boolean algebra $\mathcal{B}$ is said to be *countably decomposable* if every *disjoint* set of non-zero elements of $\mathcal{B}$ is necessarily countable. Here two elements

$b_1, b_2 \in \mathcal{B}$ are called disjoint if $b_1 \wedge b_2 = 0$ and a set $E \subseteq \mathcal{B}$ is said to be disjoint if every pair of distinct elements of E is disjoint. ∎

Exercise 54.[*] Let $2^{\mathbb{N}}$ denote the B.a. of all subsets of $\mathbb{N}$ equipped with the partial order $A \leq B$ if and only if $A \subseteq B$. Show that $2^{\mathbb{N}}$ is countably decomposable. ∎

Proposition VI.2. (a) *A (general) Boolean algebra $\mathcal{B}$ is countably decomposable if and only if every set $E \subseteq \mathcal{B}$ has a countable subset D such that D and E have the same set of upper bounds , that is, if and only if*

$$\{u \in \mathcal{B} : \ d \leq u \text{ for all } d \in D\} = \{v \in \mathcal{B} : \ e \leq v \text{ for all } e \in E\}.$$

(b) *A (general) Boolean algebra $\mathcal{B}$ which is countably decomposable and abstractly σ-complete is actually abstractly complete.*

Proof. (a) Assume first that $\mathcal{B}$ has the stated property. Let $E \subseteq \mathcal{B}$ be a disjoint set of non-zero elements. By hypothesis there is a countable set $D \subseteq E$ such that D and E have the same set of upper bounds . Suppose that $E \neq D$, in which case there exists an element $e \in E \backslash D$. By disjointness of the elements in E it follows that $e \wedge d = 0$ for all $d \in D$. So, if e' is the complement of e in $\mathcal{B}$, then $e \vee e' = 1$ and hence

$$d = d \wedge (e' \vee e) = (d \wedge e') \vee (d \wedge e) = (d \wedge e') \vee 0 = d \wedge e'.$$

Accordingly, $d \leq e'$ for all $d \in D$ (c.f. Exercise 13(f)) and so e' is an upper bound for D. But, e' is *not* an upper bound for E since $e \wedge e' = 0$ shows that $e \nleq e'$ (as $e \neq 0$ and $e \leq e'$ if and only if $e \wedge e' = e$). This is a contradiction and so no such element e can exist, that is, $E = D$. In particular, E is a countable set. This shows that $\mathcal{B}$ is countably decomposable.

Conversely, let $\mathcal{B}$ be countably decomposable. Let E be an arbitrary subset of $\mathcal{B}$. Define Λ to be the family of all elements λ in $\mathcal{B}$ such that $\lambda \leq \vee_{j=1}^{k(\lambda)} e_j$ for some finite set $\{e_1, \dots, e_{k(\lambda)}\} \subseteq E$. Note that Λ is an *ideal* in $\mathcal{B}$ (see Definition II.10) and that $E \subseteq \Lambda$ (as $e \leq e$ for any $e \in E$). So, if $v \in \mathcal{B}$ has the property that $\lambda \leq v$ for all $\lambda \in \Lambda$, then also $e \leq v$ for all $e \in E$. On the other hand if $u \in \mathcal{B}$ has the property that $e \leq u$ for all $e \in E$, then $\vee_{j=1}^{k} e_j \leq u$ for all finite subsets $\{e_1, \dots, e_k\} \subseteq E$ and so $\lambda \leq u$ for all $\lambda \in \Lambda$. This shows that E and Λ have the same set of upper bounds . Let $\mathcal{F}_\Lambda$ be the family of all subsets F of Λ such that F is disjoint and consists of non-zero elements. Partially order $\mathcal{F}_\Lambda$ by declaring $F_1 \leq F_2$ (with $F_j \in \mathcal{F}_\Lambda$) if and only if $F_1 \subseteq F_2$. If $\{F_\alpha\}$ is any totally ordered subset of $\mathcal{F}_\Lambda$, then it is easily verified that $\cup_\alpha F_\alpha \in \mathcal{F}_\Lambda$ and so $\cup_\alpha F_\alpha$ is an upper bound for $\{F_\alpha\}$ in $\mathcal{F}_\Lambda$. By *Zorn's lemma* , [14; p.6], $\mathcal{F}_\Lambda$ has a maximal element, say $\widetilde{F}$.

The claim is that $\widetilde{F}$ and Λ have the same set of upper bounds. Since $\widetilde{F} \subseteq \Lambda$ it is clear that any upper bound of Λ is also an upper bound of $\widetilde{F}$. So, suppose that $b \in \mathcal{B}$ has the property that $f \leq b$ for all $f \in \widetilde{F}$. If b is *not* an upper bound of Λ, then there exists $\lambda \in \Lambda$ such that $\lambda \nleq b$. Consider $w := \lambda \wedge b'$. Then $w \neq 0$, for otherwise (i.e. if $w = 0$)

$$\lambda \wedge b = (\lambda \wedge b) \vee w = (\lambda \wedge b) \vee (\lambda \wedge b') = \lambda \wedge (b \vee b') = \lambda \wedge 1 = \lambda,$$

which means that $\lambda \leq b$; this is a contradiction. Also, if $f \in \widetilde{F}$, then

$$f \wedge w = f \wedge b' \wedge \lambda = f \wedge b \wedge b' \wedge \lambda = 0,$$

where the second equality uses the fact that $f \wedge b = f$ (as $f \leq b$) and the last equality uses the fact that $b \wedge b' = 0$. So, w is disjoint with every element in $\widetilde{F}$. Finally, $w \notin \widetilde{F}$. Otherwise, $w := \lambda \wedge b' = f_0$ for some $f_0 \in \widetilde{F}$. Then $b \wedge (\lambda \wedge b') = b \wedge f_0$ from which it follows that $0 = f_0$ (as $b \wedge b' = 0$ and $b \wedge f_0 = f_0$). But, all elements of $\widetilde{F}$ are non-zero and so this is impossible. Moreover, since Λ is an ideal and $\lambda \in \Lambda$ we have $w = \lambda \wedge b' \in \Lambda$. But, $\widetilde{F} \cup \{w\}$ is then a set of non-zero, disjoint elements from Λ which contains $\widetilde{F}$ as a proper subset. This contradicts the maximality of $\widetilde{F}$. Accordingly, b is an upper bound for Λ.

Since $\mathcal{B}$ is countably decomposable, the set $\widetilde{F}$ is countable. Since each of the countably many elements of $\widetilde{F}$ is dominated by the supremum of some finite subset of E, it follows that the union, say D, of these finite sets is a countable subset of E with the same set of upper bounds.

(b) Every countable supremum exists in $\mathcal{B}$ by the fact that $\mathcal{B}$ is abstractly σ-complete. By part (a) the supremum of an arbitrary set exists as it coincides with some countable supremum.   ■

We have the following important consequence. First we require a definition.

Definition VI.3. Let X be a Banach space and $\mathcal{B} \subseteq \mathcal{L}(X)$ be a B.a. of projections. A vector $x \in X$ is called *separating* for $\mathcal{B}$ if $B = 0$ whenever $B \in \mathcal{B}$ satisfies $Bx = 0$.   ■

Proposition VI.3. *Let X be a separable Banach space and $\mathcal{B} \subseteq \mathcal{L}(X)$ be a Bade σ-complete B.a. of projections. Then,*

    (a) *$\mathcal{B}$ is countably decomposable,*

    (b) *$\mathcal{B}$ has a separting vector, and*

    (c) *$\mathcal{B}$ is Bade complete .*

Proof. (a) Let $\{B_\alpha\}_{\alpha \in A}$ be a disjoint set of non-zero elements from $\mathcal{B}$. Let $\{x_n\}_{n=1}^\infty$ be a countable dense set in X. Consider $\{B_\alpha\}_{\alpha \in A}$ as being a subset of $\overline{\mathcal{B}}$ (the closure of $\mathcal{B}$ in $\mathcal{L}_s(X)$, in which case $\overline{\mathcal{B}}$ is Bade complete). Then $\{B_\alpha\}_{\alpha \in A}$ is still disjoint and consists of non-zero elements in $\overline{\mathcal{B}}$. By a Zorn's lemma argument there is a maximal disjoint set of non-zero elements, say $\{\widetilde{B}_\beta\}_{\beta \in V} \subseteq \overline{\mathcal{B}}$, containing $\{B_\alpha\}_{\alpha \in A}$. Partially ordering the collection $\mathcal{F}(V)$ of all finite subsets of the index set V by inclusion, it turns out that $\sum_{\beta \in F_1} \widetilde{B}_\beta \leq \sum_{\beta \in F_2} \widetilde{B}$ (with respect to the order in $\overline{\mathcal{B}}$) whenever $F_j \in \mathcal{F}(V)$ satisfy $F_1 \subseteq F_2$. So, $\{\sum_{\beta \in F} \widetilde{B}_\beta : F \in \mathcal{F}(V)\}$ is a monotone increasing net in $\overline{\mathcal{B}}$ and hence, by the ordered convergence property of $\overline{\mathcal{B}}$, converges (in $\mathcal{L}_s(X)$) to some $B \in \overline{\mathcal{B}}$. If $B \neq I$, then $(I - B) \in \overline{\mathcal{B}}$ is a non-zero element, disjoint with every $\widetilde{B}_\beta$ which contradicts maximality. Hence, $\sum_{\beta \in V} \widetilde{B}_\beta = B = I$. Accordingly, $x_n = \Sigma_\beta \widetilde{B}_\beta x_n$, for each $n = 1, 2, \ldots$, and so there is a countable subset $V_n \subseteq V$ such that $\widetilde{B}_\beta x_n = 0$ for all $\beta \notin V_n$. By density of $\{x_n\}_{n=1}^\infty$ in X it follows that $\widetilde{B}_\beta = 0$ whenever $\beta \notin \cup_{n=1}^\infty V_n$. In particular, $B_\alpha = 0$ for every $\alpha \in A$ with $\alpha \notin \cup_{n=1}^\infty V_n$. But, $\{B_\alpha\}_{\alpha \in A}$ has *no* zero elements and so $A \subseteq \cup_{n=1}^\infty V_n$, i.e. A is a countable set.

(c) Let $\{B_\alpha\}_{\alpha \in A} \subseteq \mathcal{B}$ be an arbitrary set. Note by part (a) and Proposition VI.2 that $\mathcal{B}$ is abstractly complete . Also, by Proposition VI.2, there is a countable subset $\{B_{\alpha_n}\}_{n=1}^\infty$ of $\{B_\alpha\}_{\alpha \in A}$ such that $\vee_\alpha B_\alpha = \vee_{n=1}^\infty B_{\alpha_n}$. Then by the Bade σ-completeness of $\mathcal{B}$, we have

$$(\vee_\alpha B_\alpha)X = (\vee_{n=1}^\infty B_{\alpha_n})X = \overline{\mathrm{sp}\{\cup_n B_{\alpha_n} X\}} \subseteq \overline{\mathrm{sp}\{\cup_\alpha B_\alpha X\}}.$$

But, $B_{\alpha_0}X \subseteq (\vee_\alpha B_\alpha)X$ for each index α_0 (as $B_{\alpha_0} \leq \vee_\alpha B_\alpha$) and so $\cup_\alpha B_\alpha \subseteq (\vee_\alpha B_\alpha)X$. Since $(\vee_\alpha B_\alpha)X$ is a closed subspace of X it follows that $\overline{\mathrm{sp}(\cup_\alpha B_\alpha X)} \subseteq (\vee_\alpha B_\alpha)X$. This shows that $(\vee_\alpha B_\alpha)X = \overline{\mathrm{sp}(\cup_\alpha B_\alpha X)}$. Similarly,

$$\wedge_\alpha B_\alpha = \vee_\alpha(I - B_\alpha) = \vee_{k=1}^\infty (I - B_{\beta_k}) = \wedge_{k=1}^\infty B_{\beta_k},$$

for some countable set of indices $\{\beta_k\}_{k=1}^\infty \subseteq A$; see Proposition VI.2. Then $(\wedge_\alpha B_\alpha)X = (\wedge_{k=1}^\infty B_{\beta_k})X = \cap_{k=1}^\infty B_{\beta_k}X$ (by the Bade σ-completeness of $\mathcal{B}$). Accordingly, $\cap_\alpha B_\alpha X \subseteq \cap_{k=1}^\infty B_{\beta_k}X = (\wedge_\alpha B_\alpha)X$. But, since $\wedge_\alpha B_\alpha \leq B_{\alpha_0}$ for each index α_0, also $(\wedge_\alpha B_\alpha)X \subseteq B_{\alpha_0}X$ for every α_0. That is, $(\wedge_\alpha B_\alpha)X \subseteq \cap_\alpha B_\alpha X$. This shows that $(\wedge_\alpha B_\alpha)X = \cap_\alpha B_\alpha X$ and we have established the Bade completeness of $\mathcal{B}$.

(b) Since $\mathcal{B}$ is Bade complete, the *carrier projection* C_x of x, which is defined to be the element of $\mathcal{B}$ given by

$$C_x := \wedge\{B \in \mathcal{B} : Bx = x\},$$

exists for each $x \in X$. By a Zorn's lemma argument there is a maximal, disjoint family $\{C_{x_\alpha}\}$ of carrier projections in $\mathcal{B}$. Since $\mathcal{B}$ is countably decomposable it is of the form $\{C_{x(n)}\}_{n=1}^\infty$, where we can suppose that $\|x(n)\| = 1$, for each $n \in \mathbb{N}$ (since $C_x = C_{\lambda x}$ for all $x \in X$ and $\lambda \neq 0$ in $\mathbb{C}$). Let $\xi := \sum_{n=1}^\infty 2^{-n}x(n)$. We claim that carrier projections always satisfy the following property.

Fact. $C_x x = x$, for each $x \in X$.

To see this fix $x \in X$ and let $\mathcal{B}_x := \{B \in \mathcal{B} : Bx = x\}$. For each finite set $F \subseteq \mathcal{B}_x$ let $C_F := \wedge\{B : B \in F\}$. Given finite subsets F_1 and F_2 of $\mathcal{B}_x$ define $F_1 \leq F_2$ if $F_1 \subseteq F_2$, in which case $C_{F_2} \leq C_{F_1}$ in $\mathcal{B}$. Then $\{C_F : F \in \mathcal{F}(\mathcal{B}_x)\}$, where $\mathcal{F}(\mathcal{B}_x)$ is the family of all finite subsets of $\mathcal{B}_x$ directed by inclusion, is a decreasing net in $\mathcal{B}$ with $\wedge\{C_F : F \in \mathcal{F}(\mathcal{B}_x)\} = C_x$. Moreover, if $F = \{B_1, \dots, B_n\} \subseteq \mathcal{B}_x$, then $B_j x = x$ for each $1 \leq j \leq n$ and so

$$B_1 B_2 \dots B_n x = B_1 B_2 \dots B_{n-1}x = \dots = B_1 B_2 x = B_1 x = x.$$

That is, $C_F x = (\wedge_{j=1}^n B_j)x = B_1 B_2 \dots B_n x = x$. Since $\lim_F C_F = C_x$ in $\mathcal{L}_s(X)$ (see Theorem IV.1 and Lemma IV.1) we deduce that $C_x x = x$. This establishes the Fact.

Claim 1. $C_{x(n)}\xi = 2^{-n}x(n) = 2^{-n}C_{x(n)}x(n), \qquad n \in \mathbb{N}.$

Indeed, fix $n \in \mathbb{N}$. Since $C_{x(n)}C_{x(k)} = 0$ if $k \neq n$, it follows from the continuity of $C_{x(n)}$ and the above Fact that $C_{x(n)}\xi = \sum_{k=1}^\infty C_{x(n)}2^{-k}x(k) = \sum_{k=1}^\infty 2^{-k}C_{x(n)}C_{x(k)}x(k) = 2^{-n}C_{x(n)}x(n) = 2^{-n}x(n)$. This establishes Claim 1.

Claim 2. $\vee_{n=1}^\infty C_{x(n)} = \sum_{n=1}^\infty C_{x(n)} = I$, where the series converges in $\mathcal{L}_s(X)$.

Observe that the disjointness of $\{C_{x(n)}\}_{n=1}^\infty$ implies that $\vee_{k=1}^n C_{x(k)} = \sum_{k=1}^n C_{x(k)}$, for each $n \in \mathbb{N}$. Since the projections $\{\vee_{k=1}^n C_{x(k)}\}_{n=1}^\infty$ are increasing in the order of $\mathcal{B}$ to the projection $B := \vee_{k=1}^\infty C_{x(k)}$, it follows from Theorem IV.1 and Lemma IV.1 that

$$\lim_{n \to \infty} \sum_{k=1}^n C_{x(k)} = \lim_{n \to \infty} \vee_{k=1}^n C_{x(k)} = B, \qquad \text{in } \mathcal{L}_s(X).$$

In particular, $B = \sum_{k=1}^{\infty} C_{x(k)}$ with the series converging in $\mathcal{L}_s(X)$. Suppose that $B \neq I$, in which case $(I - B) \neq 0$. Since $C_{x(n)}B = C_{x(n)}$ (as $C_{x(n)} \leq B$) for all $n \in \mathbb{N}$, it follows that

$$(2) \qquad\qquad C_{x(n)}(I - B) = C_{x(n)}B(I - B) = 0, \qquad n \in \mathbb{N}.$$

Choose any non-zero vector $y \in (I - B)X$, in which case $(I - B)y = y$. So, by the definition of carrier projections we have $C_y \leq (I - B)$, that is, $C_y(I - B) = C_y$. It then follows from (2) that

$$C_{x(n)}C_y = C_{x(n)}C_y(I - B) = 0, \qquad n \in \mathbb{N}.$$

Accordingly, the carrier projection $C_y \neq 0$ is disjoint from $\{C_{x(n)}\}_{n=1}^{\infty}$ which contradicts the maximality of $\{C_{x(n)}\}_{n=1}^{\infty}$. Hence, $B = I$ and Claim 2 is established.

Claim 3. *The element ξ is a separating vector for $\mathcal{B}$.*

To establish this suppose that $B \in \mathcal{B}$ satisfies $B\xi = 0$. By Claim 1 and the above Fact we deduce that

$$Bx(n) = BC_{x(n)}x(n) = 2^n BC_{x(n)}\xi = 2^n C_{x(n)}B\xi = 0, \qquad n \in \mathbb{N}.$$

It follows that

$$(I - B)C_{x(n)}x(n) = (I - B)x(n) = x(n), \qquad n \in \mathbb{N},$$

and so, by the definition of carrier projections , we have that $C_{x(n)} \leq (I - B)C_{x(n)}$, that is, $C_{x(n)} \leq (C_{x(n)} - BC_{x(n)})$. It follows that $BC_{x(k)} = 0$, for all $n \in \mathbb{N}$, and so $B(\sum_{k=1}^{n} C_{x(k)}) = 0$, for all $n \in \mathbb{N}$. Let $n \to \infty$ and use Claim 2 to deduce that $0 = B(\vee_{n=1}^{\infty} C_{x(n)}) = BI = B$, that is, $B = 0$. This shows that ξ is a separating vector and completes the proof of Claim 3 and of the proposition. ∎

Exercise 55. Let X be a *separable* Banach space. Show that any spectral measure $P : \Sigma \longrightarrow \mathcal{L}_s(X)$, with Σ any σ-algebra of sets , is necessarily a *closed* spectral measure. ∎

Definition VI.4. Let X be a Banach space and $\mathcal{B} \subseteq \mathcal{L}(X)$ be a B.a. of projections .

(i) A non-zero projection $B \in \mathcal{B}$ is called an *atom* if, whenever $D \in \mathcal{B}$ satisfies $D \leq B$ then either $D = 0$ or $D = B$.

(ii) $\mathcal{B}$ is called *atomic* if there exists a family $\{B_\alpha\}_{\alpha \in A}$ of distinct atoms in $\mathcal{B}$ such that, whenever $B \in \mathcal{B}$ there is a subset $A_B \subseteq A$ such that $B = \sum_{\alpha \in A_B} B_\alpha$, that is, B is the strong operator limit of the net of finite partial sums of $\{B_\alpha : \alpha \in A_B\}$.

If the index set A in (ii) is countable, then we say that $\mathcal{B}$ is *countably atomic*. ∎

Exercise 56.[*] Let X be a Banach space and $\mathcal{B} \subseteq \mathcal{L}(X)$ be a B.a. of projections. Show that if B_1 and B_2 are distinct atoms in $\mathcal{B}$, then $B_1 \wedge B_2 = 0$ (i.e they are disjoint) . ∎

Exercise 57. Let X be a Banach space and $\mathcal{B} \subseteq \mathcal{L}(X)$ be a Bade complete B.a. of projections which is atomic. Show that the set of atoms $\{B_\alpha\}_{\alpha \in A}$ which generates $\mathcal{B}$ in the sense of Definition VI.4 is maximal and pairwise disjoint . ∎

Lemma VI.1. *Let X be a Banach space and $\mathcal{B} \subseteq \mathcal{L}(X)$ be a countably atomic , Bade σ-complete B.a. of projections. Then $\mathcal{B}$ is Bade complete .*

Proof. Let $\{B_n\}_{n=1}^{\infty}$ be a countable family of atoms in $\mathcal{B}$ which generates $\mathcal{B}$ in the sense of Definition VI.4. The map $F \mapsto \sum_{n \in F} B_n$, for each subset $F \subseteq \mathbb{N}$, is a B.a. isomorphism of the B.a. $2^{\mathbb{N}}$ (of all subsets of $\mathbb{N}$) onto $\mathcal{B}$. Since $2^{\mathbb{N}}$ is countably decomposable (see Exercise 54) so is $\mathcal{B}$. Hence, $\mathcal{B}$ is Bade complete ; this follows from an examination of the proof of Proposition VI.3(c) which *only* used the countable decomposability of $\mathcal{B}$ and not the fact that X was separable . ∎

Definition VI.5. A (general) abstractly σ-complete (resp. abstractly complete) B.a. $\mathcal{B}$ is called *countably generated* if there is a countable subset $\mathcal{D} \subseteq \mathcal{B}$ such that the smallest abstractly σ-complete (resp. abstractly complete) B.a. $\mathcal{F} \subseteq \mathcal{B}$ with $\mathcal{D} \subseteq \mathcal{F}$ is $\mathcal{F} = \mathcal{B}$. ∎

Remark. An examination of Lemma 4 in [14; p.167] and its proof, together with the fact that every abstract B.a. is isomorphic to a B.a. of subsets of some set (see Theorem II.2), shows that the countable set $\mathcal{D}$ in Definition VI.5 can always be chosen as a *Boolean subalgebra* of $\mathcal{B}$, meaning that it contains the same 0 and 1 from $\mathcal{B}$ and is closed with respect to finite sups and infs and complements. ∎

Exercise 58. Let X be the *non-separable* Hilbert space $\ell^2([0, 1])$ and, for each subset $F \subseteq [0, 1]$, let $P(F) \in \mathcal{L}(X)$ denote the projection operator in X of multiplication by χ_F. Show that $\mathcal{B} := \{P(F) : F \subseteq [0, 1]\}$ is a Bade complete B.a. of projections in $\mathcal{L}(X)$ which is *countably generated* . ∎

Lemma VI.2. *Let X be a Banach space and $\mathcal{B} \subseteq \mathcal{L}(X)$ be a Bade complete B.a. of projections with a separating vector, say x.*

(a) *$\mathcal{B}$ is countably generated if and only if its restriction to the ($\mathcal{B}$-invariant) cyclic space $\mathcal{B}[x]$ is countably generated.*

(b) *If the cyclic space $\mathcal{B}[x]$ is separable, then $\mathcal{B}$ is countably generated.*

(c) *If X is separable and x is a cyclic vector for $\mathcal{B}$, then $\mathcal{B}$ is countably generated.*

Proof. (a) Let $Y := \mathcal{B}[x]$. Since Y is invariant for each $B \in \mathcal{B}$ it follows that the family $\mathcal{B}_Y := \{B|_Y : B \in \mathcal{B}\}$ of restricted operators is Bade σ-complete in $\mathcal{L}(Y)$; see Lemma V.4. Since $\mathcal{B}_Y$ has a cyclic vector in Y (namely x), it follows from Theorem V.7 that $\mathcal{B}_Y$ is actually Bade complete . Moreover, the map $\Phi : \mathcal{B} \longrightarrow \mathcal{B}_Y$ given by $\Phi(B) = B_Y$ is a surjective B.a. homomorphism . Using the fact that x is a separating vector for $\mathcal{B}$ it is routine to check that Φ is also injective and hence, Φ is actually a B.a. isomorphism of $\mathcal{B}$ onto $\mathcal{B}_Y$. Accordingly, $\mathcal{B}$ is countably generated if and only if $\mathcal{B}_Y$ is countably generated.

(b) This follows from part (a) and part (c) below after noting that $\mathcal{B}_Y$ (with $Y := \mathcal{B}[x]$) is Bade complete; see the proof of part (a).

(c) Let $\{D_n x\}_{n=1}^{\infty}$ be a countable dense subset of the separable *subset* $W := \{Bx : B \in \mathcal{B}\}$ of X. Let $\mathcal{N}^{(\sigma)} \subseteq \mathcal{L}(X)$ be the abstractly σ-complete B.a. generated by $\{D_n\}_{n=1}^{\infty}$, in which case $\mathcal{N}^{(\sigma)} \subseteq \mathcal{B}$ of course. Let $\{B_n\}_{n=1}^{\infty}$ be a monotone sequence in $\mathcal{N}^{(\sigma)}$. Since $\mathcal{B}$ is Bade complete , it has the ordered convergence property and so $\lim_{n \to \infty} B_n = B$ exists in $\mathcal{L}_s(X)$, with $B = \vee_{n=1}^{\infty} B_n$ if $\{B_n\}$ is increasing and $B = \wedge_n B_n$ if $\{B_n\}$ is decreasing; see Theorem IV.1 and Lemma IV.1. But, since $\mathcal{N}^{(\sigma)}$ is abstractly σ-complete we have $\vee_n B_n$ and $\wedge_n B_n$ both belong to $\mathcal{N}^{(\sigma)}$. Hence, $\mathcal{N}^{(\sigma)}$ has the σ-ordered convergence property and so $\mathcal{N}^{(\sigma)}$ is Bade σ-complete by Theorem IV.1. Then Proposition VI.3 (together with the fact that any cyclic vector is necessarily a separating vector) shows that $\mathcal{N}^{(\sigma)}$ is actually Bade complete .

Fix $B \in \mathcal{B}$. Since $\mathcal{N}_x^{(\sigma)} := \{Nx : N \in \mathcal{N}^{(\sigma)}\}$ is a dense subset of W there exist elements $\{N_k\}_{k=1}^{\infty} \subseteq \mathcal{N}^{(\sigma)}$ such that $Bx = \lim_{k \to \infty} N_k x$. It follows, for every vector $z = \sum_{j=1}^{m} \alpha_j R_j x$ with $\alpha_j \in \mathbb{C}$ and $R_j \in \mathcal{B}$, that $\lim_{k \to \infty} N_k z = Bz$. Since the set of all such vectors z is dense in X and $\sup\{\|N_k\| : k \in \mathbb{N}\} < \infty$, we conclude that $N_k \longrightarrow B$ in $\mathcal{L}_s(X)$, as $k \to \infty$. This shows that B belongs to the closure $\overline{\mathcal{N}^{(\sigma)}}$, of $\mathcal{N}^{(\sigma)}$, in $\mathcal{L}_s(X)$. By the Bade completeness of $\mathcal{N}^{(\sigma)}$ and Corollary V.5.1 we deduce that $\overline{\mathcal{N}^{(\sigma)}} = \mathcal{N}^{(\sigma)}$ and so $B \in \mathcal{N}^{(\sigma)}$. Since $B \in \mathcal{B}$ is arbitrary we conclude that $\mathcal{N}^{(\sigma)} = \mathcal{B}$ and so $\mathcal{B}$ is countably generated. ∎

Remark. Proposition VI.3 and Lemma VI.2 imply that every Bade complete B.a. of projections in a separable Banach space X is necessarily countably generated. In particular, in a separable Banach space X every Bade σ-complete Boolean *subalgebra* of a Bade complete B.a. $\mathcal{B}$ (in which case $\mathcal{B}$ is necessarily countably generated) is itself countably generated. ∎

Definition VI.6. By a *measure algebra* we mean a (general) abstractly σ-complete B.a. $\mathcal{B}$ together with a function $\mu : \mathcal{B} \longrightarrow [0, \infty)$ satisfying

(i) $\mu(B) = 0$ if and only if $B = 0$,

and

(ii) $\mu(\vee_{n=1}^{\infty} B_n) = \sum_{n=1}^{\infty} \mu(B_n)$ whenever $B_n \wedge B_m = 0$ for all $m \neq n$.

In this case we call μ a measure on $\mathcal{B}$ and denote the measure algebra by $(\mathcal{B}, \mu)$. ∎

Example 26. Let Σ be a σ-algebra of subsets of some non-empty set Ω and $\nu : \Sigma \longrightarrow [0, \infty)$ be a σ-additive measure. Let $\mathcal{N} := \{E \in \Sigma : \nu(E) = 0\}$ and define $\mathcal{B} := \Sigma/\mathcal{N}$ to be the quotient B.a. of equivalence classes , where we define $E \sim F$ if $E \Delta F \in \mathcal{N}$ in which case

$$[E] := \{F \in \Sigma : E \Delta F \in \mathcal{N}\}, \qquad [E] \in \mathcal{B}.$$

If we define a function $\mu : \mathcal{B} \longrightarrow [0, \infty)$ by $\mu([E]) := \nu(E)$, then μ is well defined and $(\mathcal{B}, \mu)$ is a measure algebra. This is the classical example of a measure algebra. ∎

Let $(\mathcal{B}, \mu)$ be a measure algebra . For each pair $b_1, b_2 \in \mathcal{B}$ define

$$b_1 \Delta b_2 := (b_1 \wedge b_2') \vee (b_1' \wedge b_2).$$

Then $\rho_\mu : \mathcal{B} \times \mathcal{B} \longrightarrow [0, \infty)$ defined by $\rho_\mu(b_1, b_2) = \mu(b_1 \Delta b_2)$ specifies a *metric* in $\mathcal{B}$. It turns out that $(\mathcal{B}, \rho_\mu)$ is a *complete* metric space and that the three mappings

$$b \mapsto b', \text{ and } (b_1, b_2) \mapsto b_1 \vee b_2, \text{ and } (b_1, b_2) \mapsto b_1 \wedge b_2,$$

are continuous on $\mathcal{B}$ and $\mathcal{B} \times \mathcal{B}$ with respect to ρ_μ and $\rho_\mu \times \rho_\mu$. These facts can be found in [37; Chapter 15, Section 2], for example.

Exercise 59. Let X be a Banach space and $\mathcal{B} \subseteq \mathcal{L}(X)$ be a Bade complete B.a. of projections which is *countably generated* by some countable Boolean subalgebra $\mathcal{D} \subseteq \mathcal{B}$. Suppose that $x \in X$ is a *separating vector* for $\mathcal{B}$ and that $x' \in X'$ is a *Bade functional* for x with respect to $\mathcal{B}$. Define $\mu : \mathcal{B} \longrightarrow [0, \infty)$ by

$$\mu(B) := \langle Bx, x' \rangle, \qquad B \in \mathcal{B},$$

and let $\rho_\mu : \mathcal{B} \times \mathcal{B} \longrightarrow [0, \infty)$ be the associated metric given by

$$\rho_\mu(A, B) := \mu(A \triangle B), \qquad A, B \in \mathcal{B},$$

where $A \triangle B := (A \wedge B') \vee (A' \wedge B)$.

(a) Show that the closure $\overline{\mathcal{D}}$, of $\mathcal{D}$, in the metric space $(\mathcal{B}, \rho_\mu)$ is a Bade σ-complete B.a. of projections satisfying $\mathcal{D} \subseteq \overline{\mathcal{D}} \subseteq \mathcal{B}$.

(b) Show, moreover, that actually $\overline{\mathcal{D}} = \mathcal{B}$. ∎

Definition VI.7. Let $(\mathcal{B}, \mu)$ be a measure algebra . Then $\mathcal{B}$ is called *separable* if $(\mathcal{B}, \rho_\mu)$ is a separable metric space. ∎

The next result gives an important procedure for generating measure algebras in the setting of B.a. 's of projections.

Proposition VI.4. *Let X be a Banach space and $\mathcal{B} \subseteq \mathcal{L}(X)$ be a B.a. of projections with a separating vector $x \in X$.*

(a) *Suppose that $\mathcal{B}$ is Bade σ-complete . For any Bade functional $x' \in X'$ of x with respect to $\mathcal{B}$, define $\mu(B) := \langle Bx, x' \rangle$, for $B \in \mathcal{B}$. Then $(\mathcal{B}, \mu)$ is a measure algebra.*

(b) *Suppose, in addition to (a), that X is separable. Then the measure algebra $(\mathcal{B}, \mu)$ of part (a) is separable.*

Proof. (a) It is clear from the definition of Bade functionals that $\mu \geq 0$ is a finite-valued function on $\mathcal{B}$. Moreover, if $B \in \mathcal{B}$ satisfies $\mu(B) = 0$ (i.e. $\langle Bx, x' \rangle = 0$), then $Bx = 0$ (see Definition VI.1) and hence $B = 0$ (as x is a separating vector for $\mathcal{B}$). Suppose that $\{B_n\}_{n=1}^\infty$ is a pairwise disjoint sequence in $\mathcal{B}$. Then $\{\vee_{k=1}^n B_k\}_{n=1}^\infty$ is increasing in $\mathcal{B}$. Since $\mathcal{B}$ has the σ-monotone property we know that

$$\vee_{n=1}^\infty B_n = \lim_{n \to \infty} \vee_{k=1}^n B_k, \qquad \text{in } \mathcal{L}_w(X),$$

and hence, that

$$\mu(\vee_{n=1}^\infty B_n) = \langle (\vee_{n=1}^\infty B_n)x, x' \rangle = \lim_{n \to \infty} \langle (\vee_{k=1}^n B_k)x, x' \rangle.$$

But, by the disjointness of $\{B_n\}_{n=1}^\infty$ it follows that

$$\langle (\vee_{k=1}^n B_k)x, x' \rangle = \sum_{k=1}^n \langle B_k x, x' \rangle, \qquad n \in \mathbb{N},$$

and so

$$\mu(\vee_{n=1}^\infty B_n) = \lim_{n \to \infty} \sum_{k=1}^n \mu(B_k) = \sum_{n=1}^\infty \mu(B_n).$$

This shows that $(\mathcal{B}, \mu)$ is a measure algebra.

(b) Let us suppose now, in addition to (a), that X is separable. Proposition VI.3 shows that $\mathcal{B}$ is Bade complete and hence, by the Remark after Lemma VI.2 we see that $\mathcal{B}$ is countably generated via some countable Boolean *subalgebra* $\mathcal{D}$ of $\mathcal{B}$ (see the Remark after Definition

VI.5). It follows, since the closure of $\mathcal{D}$ in $(\mathcal{B}, \rho_\mu)$ is a Bade σ-complete B.a. containing $\mathcal{D}$ (see Exercise 59), that this closure equals $\mathcal{B}$ and hence, that $\mathcal{D}$ is dense in $\mathcal{B}$. Accordingly, $(\mathcal{B}, \mu)$ is separable. ∎

We now come to another main result of this chapter.

Theorem VI.2. *Let X be a separable Banach space and $\mathcal{B} \subseteq \mathcal{L}(X)$ be a Bade σ-complete B.a. of projections. Then there exists a scalar-type spectral operator $T \in \mathcal{L}(X)$ with $\sigma(T) \subseteq \mathbb{R}$ whose resolution of the identity $P_T : Bo(\mathbb{R}) \longrightarrow \mathcal{L}_s(X)$ satisfies $\mathcal{B} = \{P_T(\delta) : \ \delta \in Bo(\mathbb{R})\}$.*

Proof. By Proposition VI.3 $\mathcal{B}$ is Bade complete. Hence, if $\{B_\alpha\}$ is a maximal disjoint family of atoms in $\mathcal{B}$, then the series $\Sigma_\alpha B_\alpha$ is summable in $\mathcal{L}_s(X)$ to the element $B_a = \vee_\alpha B_\alpha$ of $\mathcal{B}$. Let x be a separating vector for $\mathcal{B}$ (see Proposition VI.3). Then $\Sigma_\alpha B_\alpha x = B_a x$, with the series being unconditionally norm summable in X. Accordingly, at most countably many of the vectors $\{B_\alpha x\}$ can be non-zero. Since x is a separating vector it follows that $\mathcal{B}$ has at most countably many atoms, say $\{B_n\}_{n=1}^\infty$, and hence $B_a = \vee_{n=1}^\infty B_n$. Let $X_a := B_a X$, in which case X_a is also a separable Banach space. Then the restricted B.a. $\mathcal{B}_a := \{B|_{X_a} : B \in \mathcal{B}\}$ is a Bade complete , countably atomic B.a. in X_a; see Lemma V.4 and Lemma VI.1. Let $X_c := (I - B_a)X$, in which case X_c is also a separable Banach space and we have the direct sum $X = X_a \oplus X_c$. Let $\mathcal{B}_c := \{B|_{X_c} : B \in \mathcal{B}\}$. Suppose, for the moment, that there exists a scalar-type spectral operator $T_c \in \mathcal{L}(X_c)$ with $\sigma(T_c) \subseteq \mathbb{R}$, such that the range of its resolution of the identity $P_c : Bo(\sigma(T_c)) \longrightarrow \mathcal{L}_s(X_c)$ coincides with $\mathcal{B}_c$. Then we choose an interval $[a, b] \subseteq \mathbb{R}$ having $\sigma(T_c)$ in its *interior*. Hence, $T_a := \sum_{n=1}^\infty (b + \frac{1}{n}) B_n^{(a)}$ (where $B_n^{(a)}$ is the restriction of B_n to X_a and the series converges in $\mathcal{L}_s(X_a)$) is a scalar-type spectral operator in X_a with $\sigma(T_a) = \{b\} \cup \{b + \frac{1}{n}\}_{n=1}^\infty$ contained in $\mathbb{R}$ and disjoint from $\sigma(T_c)$, and whose resolution of the identity $P_a : Bo(\sigma(T_a)) \longrightarrow \mathcal{L}_s(X_a)$ has range coinciding with $\mathcal{B}_a$.

By construction $\Lambda := \sigma(T_a) \cup \sigma(T_c)$ is a disjoint union in $\mathbb{R}$. Moreover, every Borel set $G \in Bo(\Lambda)$ has a unique decomposition $G = G_a \cup G_c$ into disjoint Borel sets $G_a \in Bo(\sigma(T_a))$ and $G_c \in Bo(\sigma(T_c))$. Define $P_T(G) \in \mathcal{L}(X)$ by $P_T(G) = P_a(G_a) \oplus P_c(G_c)$. Then the set function $P_T : Bo(\Lambda) \longrightarrow \mathcal{L}_s(X)$ so defined is a spectral measure . Moreover, if $T := T_a \oplus T_c$, in which case $\sigma(T) = \Lambda$, then $T = \int_{\sigma(T)} \lambda \, dP_T(\lambda)$ is a scalar-type spectral operator with $\sigma(T) \subseteq \mathbb{R}$ which satisfies

$$\mathcal{B} = \{P_T(G) : G \in Bo(\mathbb{R})\} = \{P_T(G) : G \in Bo(\sigma(T))\}.$$

So, it remains to exhibit an operator $T_c \in \mathcal{L}(X_c)$ with the properties required above. Now, $\mathcal{B}_c$ is a Bade complete B.a. in the separable Banach space X_c (hence is countably generated) , has *no* atoms and has $z := (I - B_a)x$ as a separating vector. Choose a Bade functional $z' \in X_c'$ such that $\langle z, z' \rangle = 1$. By Proposition VI.4 with $\mu(D) = \langle Dz, z' \rangle$, for $D \in \mathcal{B}_c$, we see that $(\mathcal{B}_c, \mu)$ is a separable measure algebra with range $\mu(\mathcal{B}_c) \subseteq [0, 1]$, satisfies $\mu(I) = 1$ and has *no atoms* . By a classical result of C. Carathéodory , [37; p.321], there is a B.a. isomorphism Φ of $\mathcal{B}_c$ onto the measure algebra $\mathcal{A}$ generated by Lebesgue measure in $[0, 1]$, which preserves countable suprema and infima. If $\mathcal{N}$ denotes the Lebesgue null sets of $Bo([0, 1])$, then $\mathcal{A}$ is B.a. isomorphic to the quotient $Bo([0, 1])/\mathcal{N}$. Let $\rho : Bo([0, 1]) \longrightarrow$

$Bo([0,1])/\mathcal{N}$ be the quotient map. Then $P_c : Bo([0,1]) \longrightarrow \mathcal{L}_s(X_c)$ given by $P_c(G) = \Phi^{-1}(\rho(G))$, for $G \in Bo([0,1])$, is a spectral measure whose range is precisely $\mathcal{B}_c$. Indeed, it is clear that $P_c(Bo([0,1])) = \mathcal{B}_c$, that P_c is multiplicative, and that P_c satisfies $P_c(\emptyset) = 0$ and $P_c([0,1]) = I$. To verify the σ-additivity of P_c let $G_n \downarrow \emptyset$ in $Bo([0,1])$. Since ρ is a surjective B.a. homomorphism, there exists $G \in Bo([0,1])$ such that $\rho(G_n) \downarrow \rho(G)$ in $Bo([0,1])/\mathcal{N}$. Let λ denote Lebesgue measure, in which case $\lambda(G_n) \downarrow 0$. Observe that λ induces a well defined action in the quotient B.a. $Bo([0,1])/\mathcal{N}$ via the formula $\lambda(\rho(E)) = \lambda(E)$, for every $E \in Bo([0,1])$. Since $\rho(G_n) \downarrow \rho(G)$, we have $\rho(G) = \rho(G)\rho(G_n) = \rho(G \cap G_n)$ for all $n \in \mathbb{N}$, and so $\lambda(G) = \lambda(G \cap G_n) \leq \lambda(G_n) \downarrow 0$. This shows that $G \in \mathcal{N}$ and so $\rho(G_n) \downarrow 0$ in the B.a. $Bo([0,1])/\mathcal{N}$. Since Φ is a B.a. isomorphism it follows that $P_c(G_n) = \Phi^{-1}(\rho(G_n)) \downarrow 0$ in the order of $\mathcal{B}_c$. But, $\mathcal{B}_c$ is Bade complete and so $P_c(G_n) \longrightarrow 0$ in $\mathcal{L}_s(X_c)$; see the equivalence (b)$\Longleftrightarrow$(c) in Theorem IV.1. This establishes the σ-additivity of P_c. Hence, $T_c := \int_{[0,1]} \lambda \, dP_c(\lambda)$ is a scalar-type spectral operator with the desired properties as $\sigma(T_c) = \mathrm{supp}(P_c) \subseteq [0,1]$. ∎

We end this chapter with an example to show that Theorem VI.2 is no longer valid without the separability assumption (even in Hilbert spaces).

Let X be a Banach space, $P : \Sigma \longrightarrow \mathcal{L}_s(X)$ be a closed spectral measure and $I_P : L^1(P) \longrightarrow \mathcal{L}_s(X)$ be its associated integration map $f \mapsto \int_\Omega f \, dP$, for $f \in L^1(P)$. If the underlying σ-algebra Σ is *countably generated as a σ-algebra* (i.e. there is a countable collection of sets $\{A_n\}_{n=1}^\infty \subseteq \Sigma$ such that the σ-algebra generated by $\{A_n\}_{n=1}^\infty$ coincides with Σ), then the locally convex Hausdorff space $(L^1(P), \tau_s(P))$ is necessarily *separable* ; see [35; Proposition 2]. Accordingly, its isomorphic image

$$I_P(L^1(P)) = \left\{ \int_\Omega f \, dP : \ f \in L^1(P) \right\}$$

must be a separable subspace of $\mathcal{L}_s(X)$ for the relative topology.

Let $T = \int_{\sigma(T)} \lambda \, dP_T(\lambda)$ be a scalar-type spectral operator (with resolution of the identity $P_T : Bo(\sigma(T)) \longrightarrow \mathcal{L}_s(X)$) such that $\sigma(T) \subseteq \mathbb{R}$. Since the σ-algebra $Bo(\sigma(T))$ is countably generated it follows from the above remarks that $I_{P_T}(L^1(P_T))$ is a separable subspace of $\mathcal{L}_s(X)$ and hence, so is its closure $\overline{I_{P_T}(L^1(P_T))}$ in $\mathcal{L}_s(X)$. Since $\overline{I_{P_T}(L^1(P_T))} = \langle \mathcal{B} \rangle_s^-$, where $\mathcal{B} := P_T(Bo(\sigma(T)))$, we have established the following result.

Proposition VI.5. *Let X be a Banach space and $\mathcal{B} \subseteq \mathcal{L}(X)$ be a Bade complete B.a. of projections. If $\mathcal{B}$ coincides with the closure (in $\mathcal{L}_s(X)$) of the resolution of the identity of some scalar-type spectral operator with real spectrum, then $\langle \mathcal{B} \rangle_s^-$ is necessarily a separable subspace of $\mathcal{L}_s(X)$.*

Example 27. Let (Ω, Σ, μ) be a finite, positive measure space such that μ is a *non-separable* measure (eg. μ can be taken to be Haar measure on the Bohr compactification Ω of the locally compact abelian group $\mathbb{R}$). Let X denote the non-separable Hilbert space $L^2(\mu)$. For each $E \in \Sigma$, let $P(E) \in \mathcal{L}(X)$ be the operator in $L^2(\mu)$ of multiplication by χ_E. Then it is routine to verify that $P : \Sigma \longrightarrow \mathcal{L}_s(X)$ so defined is a (selfadjoint) spectral measure and hence, its range $\mathcal{B} := P(\Sigma)$ is a Bade σ-complete B.a. Since the Σ-simple functions are dense in $L^2(\mu)$

it is easily verified that $1\!\!1$ (the constant function 1 on Ω) is a *cyclic vector* for $\mathcal{B}$. Then Theorem V.7 shows that $\mathcal{B}$ is actually Bade complete .

Suppose that $T = \int_{\sigma(T)} \lambda \, dP_T(\lambda)$ is a scalar-type spectral operator with $\sigma(T) \subseteq \mathbb{R}$ and satisfying $P_T(Bo(\sigma(T))) = \mathcal{B}$. The evaluation map $\Psi : S \mapsto S1\!\!1$ from $\langle \mathcal{B} \rangle_s^-$ into X is linear, continuous and its range $Y := \Psi(\langle \mathcal{B} \rangle_s^-)$ is dense in X (as $1\!\!1$ is a cyclic vector for $\mathcal{B}$). Since $\langle \mathcal{B} \rangle_s^-$ is a separable subspace of $\mathcal{L}_s(X)$– see Proposition VI.5 –it follows that Y is separable in X and hence, that $X = \overline{Y}$ is separable. This contradiction shows that no such operator T can exist.

Can there exist a resolution of the identity $P_T : Bo(\sigma(T)) \longrightarrow \mathcal{L}_s(X)$ of some scalar-type spectral operator $T \in \mathcal{L}(X)$ with $\sigma(T) \subseteq \mathbb{R}$ such that $\mathcal{B} = \overline{P_T(Bo(\sigma(T)))}$, where the closure is taken in $\mathcal{L}_s(X)$? If this were the case, then $\langle P_T(Bo(\sigma(T))) \rangle_s^- = \langle \mathcal{B} \rangle_s^-$ and so again $\langle \mathcal{B} \rangle_s^-$ would be separable, which is not the case. Hence, $\mathcal{B}$ cannot coincide with the closure (in $\mathcal{L}_s(X)$) of a resolution of the identity either.   ∎

Chapter VII

The reflexivity theorem and bicommutant algebras

The purpose of this final chapter is to establish a beautiful result, called the Bade reflexivity theorem, which states that the strongly closed operator algebra generated by a Bade σ-complete B.a. of projections $\mathcal{B} \subseteq \mathcal{L}(X)$ consists of all continuous linear operators in X which leave invariant every closed subspace of X which is invariant under every member of $\mathcal{B}$. If $\mathcal{B}$ is actually Bade complete , then this algebra also coincides with the uniformly closed operator algebra generated by $\mathcal{B}$.

The proof of this theorem is reduced to a consideration of the restricted B.a. of projections $\mathcal{B}_{\mathcal{B}[x]}$ to the cyclic spaces $\mathcal{B}[x]$, for each $x \in X$. For this special case of cyclic spaces , the proof becomes a routine consequence of an integral representation theorem characterizing the cyclic spaces $\mathcal{B}[x]$ as spaces of the type $L^1(m)$, for certain kinds of X-valued *vector measures* m. This integral representation theorem is established in the first part of the chapter. The rest of the chapter is then devoted to verifying the reflexivity theorem. As an elegant application we deduce the classical von Neumann bicommutant theorem for a bounded selfadjoint operator in a separable Hilbert space.

So, let us begin with a representation theorem of the cyclic spaces $\mathcal{B}[x]$, with $\mathcal{B}$ a Bade complete B.a. of projections, as spaces of the type $L^1(m)$ for vector measures $m : \Sigma \longrightarrow X$ of the kind $m = Px$ (with P a spectral measure satisfying $P(\Sigma) = \mathcal{B}$). For this purpose the following useful result will be needed.

Proposition VII.1. *Let X be a Banach space and $P : \Sigma \longrightarrow \mathcal{L}_s(X)$ be a spectral measure. Fix $x \in X$. Then a Σ-measurable function $f : \Omega \longrightarrow \mathbb{C}$ is Px-integrable if and only if $f \in L^1(\langle Px, x' \rangle)$, for each $x' \in X'$, and there exists a vector $x_\Omega \in X$ satisfying*

$$(1) \qquad \langle x_\Omega, x' \rangle = \int_\Omega f \, d\langle Px, x' \rangle, \qquad x' \in X'.$$

In this case, $\int_E f \, d(Px) = P(E)x_\Omega$ for each $E \in \Sigma$. In particular, f is Px-null if and only if $\int_\Omega f \, d(Px) = 0$.

Proof. If f is Px-integrable, then by definition $f \in L^1(\langle Px, x' \rangle)$, for all $x' \in X'$, and the vector $x_\Omega := \int_\Omega f \, d(Px)$ satisfies (1).

Conversely, suppose that $f \in L^1(\langle Px, x' \rangle)$, for all $x' \in X'$, and that there exists an element $x_\Omega \in X$ satisfying (1). It suffices to show that $P(E)x_\Omega$ satisfies

$$(2) \qquad \int_E f \, d\langle Px, x' \rangle = \langle P(E)x_\Omega, x' \rangle, \qquad x' \in X',$$

for each $E \in \Sigma$. So, fix $E \in \Sigma$. For each $x' \in X'$, the formula (1) implies that

$$(3) \qquad \langle P(E)x_\Omega, x' \rangle = \langle x_\Omega, P(E)'x' \rangle = \int_\Omega f \, d\langle Px, P(E)'x' \rangle.$$

Let $E_n := \{w \in \Omega : |f(w)| \le n\}$, for each $n = 1, 2, \ldots$, and $f_n := f\chi_{E_n}$. Then each function f_n belongs to $L(P)$ and so

$$(4) \qquad \int_\Omega f_n \, d\langle Px, P(E)'x' \rangle = \langle (\int_E f_n \, dP)x, x' \rangle = \int_E f_n \, d\langle Px, x' \rangle, \qquad n \in \mathbb{N}.$$

Since $|f_n| \le |f|$ and $f_n \longrightarrow f$ pointwise on Ω, the dominated convergence theorem for the complex measures $\langle Px, x' \rangle$ and $\langle Px, P(E)'x' \rangle$ yields (via (3) and (4)) that

$$\begin{aligned} \langle P(E)x_\Omega, x' \rangle &= \int_\Omega f \, d\langle Px, P(E)'x' \rangle = \lim_{n \to \infty} \int_\Omega f_n \, d\langle Px, P(E)'x' \rangle \\ &= \lim_{n \to \infty} \int_E f_n \, d\langle Px, x' \rangle = \int_E f \, d\langle Px, x' \rangle, \end{aligned}$$

which is (2). Hence, $f \in L(Px)$.

Finally, if $x_\Omega = \int_\Omega f d(P\dot{x}) = 0$, then $\int_E f \, d(Px) = P(E)x_\Omega = 0$ for all $E \in \Sigma$ and so f is Px-null. ∎

Example 28. Proposition VII.1 shows that any vector measure of the form $m = Px$, for some spectral measure $P : \Sigma \longrightarrow \mathcal{L}_s(X)$ evaluated at a vector $x \in X$, is rather *special* in that a Σ-measurable function $f : \Omega \longrightarrow \mathbb{C}$ is m-integrable if and only if

(i) $f \in L^1(\langle m, x' \rangle)$, for each $x' \in X'$,

and

(ii)   there exists $x_\Omega \in X$ satisfying $\langle x_\Omega, x' \rangle = \int_\Omega f \, d\langle m, x' \rangle$, for all $x' \in X'$.

Of course, if the Banach space X does not contain an isomorphic copy of c_0, then (i) alone suffices for f to be m-integrable ; see Exercise 42. However, for an arbitrary Banach space X and an arbitrary vector measure $m : \Sigma \longrightarrow X$ conditions (i) and (ii) do *not suffice* for f to be m-integrable.

To see this let $X := c_0$. Let $\Omega := \mathbb{N}$ and $\Sigma := 2^\Omega$. Define $m : \Sigma \longrightarrow X$ by

$$m(E) = ((\delta_1 - \delta_2)(E), 3^{-1}(\delta_3 - \delta_4)(E), 5^{-1}(\delta_5 - \delta_6)(E), \ldots), \qquad E \in \Sigma,$$

where δ_n is the Dirac point measure at $n \in \Omega$. If $\xi = (\xi_1, \xi_2, \ldots)$ belongs to $X' = \ell^1$, then it is routine to verify that

$$\langle m, \xi \rangle(E) = \xi_1 \cdot (\delta_1 - \delta_2)(E) + 3^{-1}\xi_2 \cdot (\delta_3 - \delta_4)(E) + 5^{-1}\xi_3 \cdot (\delta_5 - \delta_6)(E) + \ldots, \qquad E \in \Sigma,$$

from which it is clear that $\langle m, \xi \rangle$ is a complex measure . So, Proposition I.1 implies that m is a c_0-valued vector measure.

Now, the function $f : \Omega \longrightarrow \mathbb{C}$ defined by $f(n) = n = f(n+1)$, for each $n \in \{1, 3, 5, \ldots\}$, clearly satisfies (i) since, for each $\xi \in \ell^1$, we have

$(*) \quad \int_E f \, d\langle m, \xi \rangle = \xi_1 \cdot (\delta_1 - \delta_2)(E) + \xi_2 \cdot (\delta_3 - \delta_4)(E) + \xi_3 \cdot (\delta_5 - \delta_6)(E) + \ldots , \qquad E \in \Sigma.$

In particular, $\int_\Omega |f| \, d|\langle m, \xi \rangle| \leq 2\|\xi\|_1 < \infty$. Moreover, the element $x_\Omega := 0$ of X clearly satisfies (ii). However, for an arbitrary set $E \in \Sigma$ it can be seen from $(*)$ that only the vector

$$x_E := ((\delta_1 - \delta_2)(E), \; (\delta_3 - \delta_4)(E), \; (\delta_5 - \delta_6)(E), \ldots)$$

satisfies $\langle x_E, \xi \rangle = \int_E f \, d\langle m, \xi \rangle$ for all $\xi \in X' = \ell^1$. But, x_E is typically only an element of the bidual $X'' = \ell^\infty$ (unless E is finite or cofinite) and not an element of $X = c_0$ itself. Accordingly, f is *not* m-integrable (in X). $\blacksquare$

We can now establish the following important representation theorem.

Theorem VII.1. *Let X be a Banach space and $\mathcal{B} \subseteq \mathcal{L}(X)$ be a B.a. of projections which coincides with the range of some closed spectral measure $P : \Sigma \longrightarrow \mathcal{L}_s(X)$. Then, for each $x \in X$, the integration map $I_{Px} : L^1(Px) \longrightarrow X$ given by*

$$I_{Px} : f \mapsto \int_\Omega f \, d(Px), \qquad f \in L^1(Px),$$

is a Banach space isomorphism from $L^1(Px)$ onto the cyclic space $\mathcal{B}[x]$.

Proof. Proposition VII.1 implies that I_{Px} is injective .

Fix $x \in X$. Suppose that $f \geq 0$ is Px-integrable . Choose functions $s_n \in \mathrm{sim}(\Sigma)$, for $n \in \mathbb{N}$, such that $0 \leq s_n \uparrow f$ pointwise on Ω. By the dominated convergence theorem for the X-valued vector measure Px (see Theorem I.9) we have that $\int_\Omega s_n \, d(Px) \longrightarrow \int_\Omega f \, d(Px) = I_{Px}(f)$ in X. Since clearly $\int_\Omega s_n \, d(Px) \in \mathcal{B}[x]$, for each $n \in \mathbb{N}$, and $\mathcal{B}[x]$ is closed, it follows that $I_{Px}(f) \in \mathcal{B}[x]$ for all non-negative $f \in L^1(Px)$ and hence, for arbitrary $f \in L^1(Px)$. So, I_{Px} takes all of its values in $\mathcal{B}[x]$.

If f is Px-integrable, then we know from Proposition I.2 applied to $m := Px$ that

$$\|I_{Px}(f)\| = \Big\| \int_\Omega f \, d(Px) \Big\| \leq \|(Px)_f\|(\Omega) = \|f\|_{L^1(Px)}$$

and also that

$$\|f\|_{L^1(Px)} \;\leq\; 4 \sup\Big\{ \Big\| \int_E f \, d(Px) \Big\| : \; E \in \Sigma \Big\} = 4 \sup\Big\{ \Big\| P(E) \int_\Omega f \, d(Px) \Big\| : E \in \Sigma \Big\}$$

$$\leq\; 4\|\mathcal{B}\| \cdot \Big\| \int_\Omega f \, d(Px) \Big\| = 4\|\mathcal{B}\| \cdot \|I_{Px}(f)\|.$$

That is, for each $f \in L^1(Px)$, we have

(5) $$\|I_{Px}(f)\| \leq \|f\|_{L^1(Px)} \leq 4\|\mathcal{B}\| \cdot \|I_{Px}(f)\|,$$

which shows that I_{Px} is an isomorphism of $L^1(Px)$ onto its range $I_{Px}(L^1(Px)) \subseteq \mathcal{B}[x]$.

So, it remains to check that I_{Px} is *onto* $\mathcal{B}[x]$. Let $z \in \mathcal{B}[x]$. Then $z = \lim_{n \to \infty} z_n$ with each z_n being some finite linear combination of elements from $\{P(E)x : E \in \Sigma\}$. It is then possible to write $z_n = \sum_{j=1}^{k_n} \alpha_j^{(n)} P(E_j^{(n)})$, with $\alpha_j^{(n)} \in \mathbb{C}$ and the sets $\{E_j^{(n)}\}_{j=1}^{k_n} \subseteq \Sigma$ pairwise disjoint, so that $z_n = I_{Px}(s_n)$ with $s_n := \sum_{j=1}^{k_n} \alpha_j^{(n)} \chi_{E_j^{(n)}} \in \mathrm{sim}(\Sigma)$. Hence, $z = \lim_{n \to \infty} I_{Px}(s_n)$. It is clear from (5) that $\{s_n\}_{n=1}^{\infty}$ is then a Cauchy sequence in $L^1(Px)$. Since $L^1(Px)$ is complete (see Theorem I.11), there is $f \in L^1(Px)$ such that $s_n \longrightarrow f$ in $L^1(Px)$. By (5) again it follows that $\lim_{n \to \infty} I_{Px}(s_n) = I_{Px}(f)$. Accordingly, $z = I_{Px}(f)$. ∎

The following fact is a useful consequence of Theorem VII.1.

Exercise 60. Let X be a Banach space and $\mathcal{B} \subseteq \mathcal{L}(X)$ be a Bade σ-complete B.a. of projections with a cyclic vector $x \in X$ (i.e. $X = \mathcal{B}[x]$). Let $x' \in X'$ be a Bade functional for x with respect to $\mathcal{B}$ and let $Z \subseteq X'$ be the linear span of $\{B'x' : B \in \mathcal{B}\}$. Show that Z is dense in X' for the weak-star topology $\sigma(X', X)$. ∎

We now turn towards the main topic of this chapter.

Let $\mathcal{A} \subseteq \mathcal{L}(X)$ be a *commuting* family of operators. By $\mathrm{Lat}(\mathcal{A})$ we denote the family of all closed subspaces of X which are invariant for every operator $A \in \mathcal{A}$. We then define

$$\mathrm{AlgLat}(\mathcal{A}) := \{T \in \mathcal{L}(X) : TY \subseteq Y \text{ for all } Y \in \mathrm{Lat}(\mathcal{A})\}.$$

Exercise 61. Let X be a Banach space and $\mathcal{B} \subseteq \mathcal{L}(X)$ be a bounded B.a. of projections. Let $\overline{\mathcal{B}}$ denote the closure of $\mathcal{B}$ in $\mathcal{L}_s(X)$.

(a) Show that $\mathrm{Lat}(\mathcal{B}) = \mathrm{Lat}(\overline{\mathcal{B}})$.

(b) Show that $T \in \mathcal{L}(X)$ belongs to $\mathrm{AlgLat}(\mathcal{B})$ if and only if $TB[x] \subseteq \mathcal{B}[x]$, for each $x \in X$, where $\mathcal{B}[x]$ is the cyclic space generated by x with respect to $\mathcal{B}$.

(c) Show that $\mathrm{AlgLat}(\mathcal{B})$ is a subalgebra of $\mathcal{L}(X)$ which is closed in each of the spaces $\mathcal{L}_w(X)$, $\mathcal{L}_s(X)$ and $\mathcal{L}_u(X)$, and that $\langle \mathcal{B} \rangle_s^- \subseteq \mathrm{AlgLat}(\mathcal{B})$.

(d) Show that $\mathrm{Lat}(\mathcal{B}) = \mathrm{Lat}(\langle \mathcal{B} \rangle_s^-)$.

(e) Show that $\mathrm{AlgLat}(\mathcal{B}) = \mathrm{AlgLat}(\langle \mathcal{B} \rangle_s^-)$. ∎

Let $\mathcal{A} \subseteq \mathcal{L}(X)$ be a family of operators. Then the *commutant* $\mathcal{A}^c$, of $\mathcal{A}$, is defined by

$$\mathcal{A}^c := \{T \in \mathcal{L}(X) : TA = AT \text{ for all } A \in \mathcal{A}\}.$$

The family of operators $(\mathcal{A}^c)^c$, denoted simply by $\mathcal{A}^{cc}$, is called the *bicommutant* of $\mathcal{A}$. Of course, we have that

$$\mathcal{A}^{cc} = \{T \in \mathcal{L}(X) : TS = ST \text{ for all } S \in \mathcal{A}^c\}.$$

Exercise 62. Let X be a Banach space and $\mathcal{B} \subseteq \mathcal{L}(X)$ be a B.a. of projections.

(a) Show that $\mathcal{B}^c$ is a subalgebra of $\mathcal{L}(X)$ and that it is closed in each of the spaces $\mathcal{L}_w(X)$, $\mathcal{L}_s(X)$ and $\mathcal{L}_u(X)$. Show that each of the subalgebras $\langle \mathcal{B} \rangle_u^-$ and $\langle \mathcal{B} \rangle_s^- = \langle \mathcal{B} \rangle_w^-$ is contained in $\mathcal{B}^c$.

(b) Show that $\mathcal{B}^{cc}$ is a *commutative* subalgebra of $\mathcal{L}(X)$ which contains $\mathcal{B}$ and that it is closed in each of the spaces $\mathcal{L}_w(X)$, $\mathcal{L}_s(X)$ and $\mathcal{L}_u(X)$.

(c) Give an example of a Banach space X and a B.a. $\mathcal{B} \subseteq \mathcal{L}(X)$ such that $\mathcal{B}^c$ is *not* commutative.

(d) Let $\mathcal{B}$ be bounded and $\overline{\mathcal{B}}$ be the closure of $\mathcal{B}$ in $\mathcal{L}_s(X)$. Show that $\mathcal{B}^c = (\overline{\mathcal{B}})^c$ and $\mathcal{B}^{cc} = (\overline{\mathcal{B}})^{cc}$. ∎

The main aim of this chapter is to establish the Bade reflexivity theorem. We begin with the following weakened form of this result.

Proposition VII.2. *Let X be a Banach space and $\mathcal{B} \subseteq \mathcal{L}(X)$ be a Bade σ-complete B.a. of projections with a cyclic vector. Then*

$$(6) \qquad \langle \mathcal{B} \rangle_u^- = \mathcal{B}^c.$$

Proof. It is routine to verify that $\langle \mathcal{B} \rangle_u^- \subseteq \mathcal{B}^c$; see Exercise 62(a).

To prove the converse, let $T \in \mathcal{B}^c$. Let $P : \Sigma \longrightarrow \mathcal{L}_s(X)$ be a spectral measure such that $P(\Sigma) = \mathcal{B}$. Since $\mathcal{B}$ is Bade complete (c.f. Theorem V.7) it follows that P is a *closed* spectral measure (see Theorem IV.1) and hence, by Theorem VII.1, that the integration map $I_{Px} : f \mapsto \int_\Omega f \, d(Px)$ is a bicontinuous isomorphism of $L^1(Px)$ onto $X = \mathcal{B}[x]$; here $x \in X$ is any cyclic vector for $\mathcal{B}$. Let $f \in L^1(Px)$ satisfy $Tx = \int_\Omega f \, d(Px)$ and define sets $E_n := \{w \in \Omega : |f(w)| \leq n\}$, for $n = 1, 2, \ldots$ Then $f\chi_{E_n} \in L^\infty(P)$ and so $T_n := \int_\Omega f\chi_{E_n} \, dP \in \langle \mathcal{B} \rangle_u^-$ for each $n = 1, 2, \ldots$; see the proof of Theorem V.2(a). It follows from Proposition VII.1 that

$$(7) \qquad P(E_n)Tx = \int_\Omega f\chi_{E_n} \, d(Px) = (\int_\Omega f\chi_{E_n} \, dP)x = T_n x, \qquad n \in \mathbb{N}.$$

Using (7) and the fact that $TP(E) = P(E)T$, for all $E \in \Sigma$, we deduce that $P(E_n)Tz = T_n z$, for each $n \in \mathbb{N}$ and every $z \in \text{sp}\{P(E)x : E \in \Sigma\}$. Since the space $\text{sp}\{P(E)x : E \in \Sigma\}$ is dense in X (as x is a cyclic vector), it follows that $P(E_n)T = T_n$, for each $n \in \mathbb{N}$. Noting that $E_n \uparrow \Omega$ it follows from the σ-additivity of P that $\lim_{n \to \infty} T_n = T$ in $\mathcal{L}_s(X)$ and hence, $T \in \langle \mathcal{B} \rangle_s^- = \langle P(\Sigma) \rangle_u^-$. Since $\langle \mathcal{B} \rangle_s^- = \langle \mathcal{B} \rangle_u^-$ (c.f. Corollary V.6.3), we see that $T \in \langle \mathcal{B} \rangle_u^-$. ∎

Exercise 63. Let X be a Banach space and $\mathcal{B} \subseteq \mathcal{L}(X)$ be a Bade σ-complete B.a. of projections with a cyclic vector . Show that $\mathcal{B}^c$ is necessarily commutative. ∎

B.a.'s of projections $\mathcal{B}$ which satisfy (6) are rather special. There are many known criteria on $\mathcal{B}$, other than those imposed by the hypotheses of Proposition VII.2, which also imply (6). We will only discuss one such set of criteria arising in connection with a notion of completeness for B.a.'s of projections (due to C.Rall, [33]) which is closely related to the existence of carrier projections and to Bade completeness.

Definition VII.1. Let X be a Banach space. A bounded B.a. of projections $\mathcal{B} \subseteq \mathcal{L}(X)$ is called τ-*complete* if, for each $x \in X$, there exists a *smallest* projection $P \in \mathcal{B}$ (depending on x) satisfying $Px = x$. That is, whenever $Q \in \mathcal{B}$ satisfies $Qx = x$, then necessarily $P \leq Q$. ∎

Example 29. Any Bade complete B.a. of projections $\mathcal{B} \subseteq \mathcal{L}(X)$ is τ-complete. Indeed, for each $x \in X$ the carrier projection $C_x := \wedge\{B \in \mathcal{B} : Bx = x\}$ exists in $\mathcal{B}$ (by abstract completeness) and it necessarily satisfies $C_x x = x$ (see the Fact in the proof of Proposition VI. 3(b)). ∎

In view of Example 29 it is natural to ask whether every abstractly complete B.a. of projections is τ-complete ? As shown by the following example this is false in general, even if $\mathcal{B}$ has a cyclic vector !

Example 30.   Let $\mathbb{N}$ be equipped with its discrete topology and let $\Omega := \beta(\mathbb{N})$. Equip the subset $\Lambda := \Omega\backslash\mathbb{N}$ with the relative topology from Ω and consider the Banach space $X :=$ $X_1 \oplus X_2 := \{(f_1, f_2) : f_j \in X_j\}$ equipped with the norm $\|(f_1, f_2)\| = \max\{\|f_1\|_\infty, \|f_2\|_\infty\}$, where $X_1 := \{f \in C(\Omega) : f\chi_\Lambda = 0\}$ and $X_2 := C_b(\Lambda)$ is the Banach space of all bounded continuous functions on Λ. We note that X_1 is Banach space isomorphic to c_0 and that X can be identified with a subspace of $\ell^\infty(\Omega)$. Since $\Lambda \cap E \in Co(\Lambda)$ whenever $E \in Co(\Omega)$, we can define a projection $P(E) \in \mathcal{L}(X)$ by

$$P(E)(f_1, f_2) = (\chi_E f_1, \chi_{E\cap\Lambda} f_2), \qquad (f_1, f_2) \in X,$$

for each $E \in Co(\Omega)$. If $\mathcal{B} := \{P(E) : E \in Co(\Omega)\}$, then it is routine to verify that $\|\mathcal{B}\| = 1$ and that $P : Co(\Omega) \longrightarrow \mathcal{B}$ is a B.a. homomorphism of $Co(\Omega)$ onto $\mathcal{B}$. To see P is injective suppose that $P(E) = 0$ for some $E \in Co(\Omega)$. Then, for each fixed $n \in \mathbb{N}$, we have $P(E)(\chi_{\{n\}}, 0) = (0, 0)$, that is, $\chi_{E\cap\{n\}} = 0$ in X_1. So, $\chi_E(n) = 0$ for each $n \in \mathbb{N}$. But, χ_E is continuous on Ω and $\mathbb{N}$ is dense in Ω from which it follows that $\chi_E = 0$ in $C(\Omega)$, i.e. $E = \emptyset$. Hence, P is actually a B.a. isomorphism of $Co(\Omega)$ onto $\mathcal{B}$. Since Ω is compact and extremely disconnected, it follows that the B.a. $Co(\Omega)$ is abstractly complete (c.f. Example 13). Hence, its isomorphic image $\mathcal{B}$ is also abstractly complete (c.f. Exercise 18).

Let $e_1 \in X_1$ be the function on Ω given by $e_1 := \sum_{n=1}^\infty \frac{1}{n}\chi_{\{n\}}$ and $e_2 := \chi_\Lambda \in X_2$, in which case $e := (e_1, e_2) \in X$. Then the cyclic space $\mathcal{B}[e] \subseteq X$ is generated by

$$\{P(E)e : E \in Co(\Omega)\} = \{(\chi_E e_1, \chi_{E\cap\Lambda}) : E \in Co(\Omega)\}.$$

By considering the sets $\{n\} \in Co(\Omega)$, for each $n \in \mathbb{N}$, it follows that $\{(\chi_{\{n\}}, 0) : n \in \mathbb{N}\} \subseteq \mathcal{B}[e]$ and hence that $X_1 \oplus \{0\} \subseteq \mathcal{B}[e]$. Given any $E \in Co(\Omega)$, the element $P(E)e = (\chi_E e_1, 0) \oplus (0, \chi_{E\cap\Lambda} e_2)$ belongs to $\mathcal{B}[e]$. Since also $\chi_E e_1 \in X_1$ (i.e. $(\chi_E e_1, 0) \in \mathcal{B}[e]$), it follows that $(0, \chi_{E\cap\Lambda} e_2) = (0, \chi_{E\cap\Lambda}) \in \mathcal{B}[e]$. Since $E \in Co(\Omega)$ is arbitrary it follows that $\{0\} \oplus X_2 \subseteq \mathcal{B}[e]$. Hence, $\mathcal{B}[e] = X$ and so e is a cyclic vector for $\mathcal{B}$.

To see that $\mathcal{B}$ is *not* τ-complete, let $E_n := \Omega\backslash\{n\}$ and $x := (0, e_2)$, and note that $P(E_n)x = x$ for all $n \in \mathbb{N}$. The claim is that $\wedge_n P(E_n) = 0_\mathcal{B}$. To establish this it suffices to show (by taking complements and using Exercise 16(b)) that $\vee_n P(\{n\}) = I$. So, let $F \in Co(\Omega)$ be any set such that $P(\{n\}) \leq P(F)$ for all $n \in \mathbb{N}$. Since P is a B.a. isomorphism, it follows that $\{n\} \subseteq F$ for all $n \in \mathbb{N}$, i.e. $\mathbb{N} \subseteq F$. But, F is closed and so also $\overline{\mathbb{N}} = \Omega \subseteq F$. Hence, $F = \Omega$ and so $P(F) = I$ or, equivalently, $\wedge_n P(E_n) = 0_\mathcal{B}$. In particular, $(\wedge_n P(E_n))x = 0$ and so $(\wedge_n P(E_n))x \neq x$. Accordingly, there is no smallest projection P in $\mathcal{B}$ satisfying $Px = x$. That is, $\mathcal{B}$ is not τ-complete .   ∎

In contrast to Example 30 we now record some positive facts about τ-completeness in relation to Bade completeness ; all of the following facts (and the terminology used below) can be found in or follow immediately from the results in.[33]. In particular, (ii) of part (b) of the following theorem is a version of (6).

Theorem VII.2. (a) *Let X be a Banach lattice with σ-order continuous norm (i.e. X is Dedekind σ-complete) and $\mathcal{B} \subseteq \mathcal{L}(X)$ be the (necessarily bounded) B.a. of all band projections in X. Then $\mathcal{B}$ is τ-complete .*

(b) *Let X be a Banach space and $\mathcal{B} \subseteq \mathcal{L}(X)$ be a bounded B.a. of projections which is τ-complete and has a cyclic vector. Then,*

(i) $\mathcal{B}$ *is abstractly σ-complete,*

and

(ii) $\langle \mathcal{B} \rangle_u^- = \langle \mathcal{B} \rangle_s^- = \text{AlgLat}(\mathcal{B}) = \mathcal{B}^c$.

(c) *Let X be a Banach space and $\mathcal{B} \subseteq \mathcal{L}(X)$ be a bounded B.a. of projections with a cyclic vector. Then the following statements are equivalent.*

(i) $\mathcal{B}$ *is Bade σ-complete .*

(ii) $\mathcal{B}$ *is Bade complete .*

(iii) $\mathcal{B}$ *is τ-complete and* $\text{Lat}(\mathcal{B}) = \{BX : B \in \mathcal{B}\}$.

Remark. (1) Part (a) of Theorem VII.2 is false in general if the Banach lattice X is not Dedekind σ-complete. Let $X := c$ be the closed subspace of ℓ^∞ consisting of all elements $x = (x_1, x_2, \ldots)$ for which $\lim_{n \to \infty} x_n$ exists. Then X is a Banach lattice (with respect to the order defined co-ordinatewise) which is not Dedekind σ-complete. Let Σ denote the algebra of subsets of $\mathbb{N}$ which are finite or have finite complement and, for each $E \in \Sigma$, define a projection $P(E) \in \mathcal{L}(X)$ by

$$P(E)x = (\chi_E(1)x_1, \chi_E(2)x_2, \ldots), \qquad x = (x_1, x_2, \ldots) \in X.$$

Then $\mathcal{B} := \{P(E) : E \in \Sigma\}$ is the B.a. of all band projections in X. To see that $\mathcal{B}$ is not τ-complete, let $x = (1, 0, \frac{1}{3}, 0, \frac{1}{5}, 0, \ldots)$. If $E \in \Sigma$ satisfies $P(E)x = x$, that is,

$$\left(\chi_E(1), 0, \frac{1}{3}\chi_E(3), 0, \frac{1}{5}\chi_E(5), 0, \ldots\right) = \left(1, 0, \frac{1}{3}, 0, \frac{1}{5}, 0, \ldots\right)$$

then we see that $n \in E$ for all odd integers n from $\mathbb{N}$. In particular, E is then an infinite set and so E^c must be finite (as $E \in \Sigma$). Hence, there must exist some even integer $n_0 \in E$ in which case $E\backslash\{n_0\} \in \Sigma$ and so $P(E\backslash\{n_0\}) \in \mathcal{B}$. But, $P(E\backslash\{n_0\})x = x$ and $P(E\backslash\{n_0\}) \leq P(E)$ with $P(E\backslash\{n_0\}) \neq P(E)$. So, there can be no smallest projection $P(E) \in \mathcal{B}$ satisfying $P(E)x = x$. In particular, $\mathcal{B}$ is not τ-complete .

It may be interesting to note that $e := (1, 1, \ldots) \in X$ is a cyclic vector for $\mathcal{B}$. It is also not difficult to verify that $\mathcal{B}$ is a closed subset of $\mathcal{L}_s(X)$, but $\mathcal{B}$ is not abstractly σ-complete.

(2) Let X be a Dedekind σ-complete Banach lattice and $\mathcal{B} \subseteq \mathcal{L}(X)$ be the B.a. of all band projections . If $\mathcal{B}$ has a cyclic vector, then it follows from parts (a) and (c) of Theorem VII.2 that $\mathcal{B}$ is Bade complete if and only if $\text{Lat}(\mathcal{B}) = \{BX : B \in \mathcal{B}\}$. In particular, if $\mathcal{B}$ is known *not* to be Bade complete (or Bade σ-complete), then the inclusion $\{BX : B \in \mathcal{B}\} \subseteq \text{Lat}(\mathcal{B})$ is necessarily strict. ∎

We now return to the main result of this chapter; its proof is an elaboration of that given in [15; Theorem 16, p.2209].

Theorem VII.3. *Let X be a Banach space and $\mathcal{B} \subseteq \mathcal{L}(X)$ be a Bade complete B.a. of projections. Then*

$$\langle \mathcal{B} \rangle_u^- = \text{AlgLat}(\mathcal{B}).$$

Proof. The inclusion $\langle \mathcal{B} \rangle_u^- \subseteq \text{AlgLat}(\mathcal{B})$ is clear; see Exercise 60(c).

For the converse, let $\Omega_\mathcal{B}$ be the Stone space of $\mathcal{B}$. By Theorem V.3(b) there is a bicontinuous isomorphism $\Phi : C(\Omega_\mathcal{B}) \longrightarrow \langle \mathcal{B} \rangle_u^-$ given by

$$\Phi(f) := \int_{\Omega_\mathcal{B}} f\, dP, \qquad f \in C(\Omega_\mathcal{B}),$$

with $P : Bo(\Omega_\mathcal{B}) \longrightarrow \mathcal{L}_s(X)$ a regular, closed, σ-additive spectral measure satisfying $P(Bo(\Omega_\mathcal{B})) = \mathcal{B}$ and $P(E) = Q(E)$, for $E \in Co(\Omega_\mathcal{B})$, where $Q : Co(\Omega_\mathcal{B}) \longrightarrow \mathcal{B}$ is the Stone map. Thus, each projection $B \in \mathcal{B}$ determines a characteristic function $\chi_{\delta(B)}$, for some clopen set $\delta(B) \in Co(\Omega_\mathcal{B})$ (i.e. $\chi_{\delta(B)} \in C(\Omega_\mathcal{B})$), such that $B = \Phi(\chi_{\delta(B)}) = P(\delta(B))$. Since $\Omega_\mathcal{B}$ is totally disconnected (actually, extremely disconnected) the sets from $Co(\Omega_\mathcal{B})$ form a base for the topology of $\Omega_\mathcal{B}$.

Since $\mathcal{B}$ is Bade complete , the carrier projections

$$(8) \qquad\qquad C_x := \wedge\{B \in \mathcal{B} :\ Bx = x\}, \qquad x \in X,$$

exist in $\mathcal{B}$; the corresponding clopen set $\delta(C_x)$ will be denoted simply by δ_x. Moreover, by the Fact in the proof of Proposition VI.3(b), we have

$$(9) \qquad\qquad x = C_x x, \qquad x \in X.$$

The next claim is (for a fixed $x \in X$) that:

$$(10) \qquad f \in C(\Omega_\mathcal{B}) \text{ with both } f\chi_{\Omega_\mathcal{B} \setminus \delta_x} = 0 \text{ and } \Phi(f)x = 0 \text{ implies that } f = 0.$$

To see this, suppose that $f \neq 0$ (under the given hypotheses). Since $Co(\Omega_\mathcal{B})$ forms a base for the topology in $\Omega_\mathcal{B}$ and $f^{-1}(\mathbb{C}\backslash\{0\})$ is an open set which is non-empty, there exists $B \neq 0$ in $\mathcal{B}$ such that $f(w) \neq 0$ for every $w \in \delta(B)$. Since $\delta(B)$ is clopen it is compact and so $g := (1/f)\cdot\chi_{\delta(B)} \in C(\Omega_\mathcal{B})$. By the fact that Φ is a homomorphism we deduce that $\Phi(g)\Phi(f) = \Phi(\chi_{\delta(B)}) = B$. Since $f\chi_{\Omega_\mathcal{B} \setminus \delta_x} = 0$, we have $f = f\chi_{\delta_x}$ and so $\Phi(f) = \Phi(f)\Phi(\chi_{\delta_x}) = \Phi(f)C_x$. Hence, multiplying by $\Phi(g)$ and using $\Phi(g)\Phi(f) = B$ gives $B = BC_x$, that is, $B \leq C_x$. So, $(C_x - B)$ is a projection in $\mathcal{B}$; see Exercise 30. Now, $Bx = \Phi(g)\Phi(f)x = 0$ (as $\Phi(f)x = 0$) and so (9) yields that $(C_x - B)x = x$. By (8) we then have that $C_x \leq (C_x - B)$. Hence, $C_x(C_x - B) = C_x$ which yields $C_x - C_x B = C_x$, that is, $C_x B = 0$. But, we saw above that $B = BC_x(= C_x B)$ and so we conclude that $B = 0$ which is a contradiction to the choice of B. This establishes (10).

Let $T \in \text{AlgLat}(\mathcal{B})$. If $B \in \mathcal{B}$, then the range $BX \in \text{Lat}(\mathcal{B})$ and the range $(I - B)X \in \text{Lat}(\mathcal{B})$ and so $TBX \subseteq BX$ and $T(I - B)X \subseteq (I - B)X$. Hence, if $x \in X$, then $TBx \in BX$

and so $B(TBx) = TBx$. Also $T(I - B)x \in (I - B)X$ and so $B[T(I - B)x] = 0$ (as $B(I - B) = 0$). Since $x \in X$ is arbitrary, we deduce that $BTB = TB$ and $BT(I - B) = 0$, from which it follows that

$$TB = BTB = BTB + BT(I - B) = BT.$$

So, whenever $T \in \mathrm{AlgLat}(\mathcal{B})$ we have established that

(11) $$TB = BT, \qquad B \in \mathcal{B}.$$

Fix now $T \in \mathrm{AlgLat}(\mathcal{B})$. By Exercise 61(b) we have that $T\mathcal{B}[x] \subseteq \mathcal{B}[x]$, for each $x \in X$. Fix $x \in X$. Then the restriction $T|_{\mathcal{B}[x]}$ of T to the cyclic space $\mathcal{B}[x]$ commutes with the restricted spectral measure $P_{\mathcal{B}[x]} : E \mapsto P(E)|_{\mathcal{B}[x]}$; see (11). An examination of the proof of Proposition VII.2 (applied in the space $\mathcal{B}[x]$ to $P_{\mathcal{B}[x]}(\Sigma) \subseteq \mathcal{L}(\mathcal{B}[x]))$ shows that $T|_{\mathcal{B}[x]} = \int_\Omega g_x \, dP_{\mathcal{B}[x]}$ for some $g_x \in L^\infty(P_{\mathcal{B}[x]})$. Choose an everywhere defined, bounded $Bo(\Omega_\mathcal{B})$-measurable function h_x such that $\sup\{|h_x(w)| : w \in \Omega_\mathcal{B}\} = |g_x|_{P_{\mathcal{B}[x]}}$ and $h_x = g_x$, $P_{\mathcal{B}[x]}$-a.e., in which case

$$T|_{\mathcal{B}[x]} = \int_\Omega h_x \, dP_{\mathcal{B}[x]}.$$

Then Theorem V.1 shows that

$$\sup_{w \in \Omega_\mathcal{B}} |h_x(w)| = |g_x|_{P_{\mathcal{B}[x]}} \leq \|T|_{\mathcal{B}[x]}\| \leq \|T\|.$$

If we define $T_x \in \mathcal{L}(X)$ by

(12) $$T_x := C_x \left(\int_{\Omega_\mathcal{B}} h_x \, dP \right), \qquad x \in X,$$

then clearly $T_x \in \langle \mathcal{B} \rangle_u^-$, as $C_x \in \mathcal{B} = P(Bo(\Omega_\mathcal{B}))$ and $h_x \in L^\infty(P)$. Since $C_x x = x$ and $(\int_{\Omega_\mathcal{B}} h_x \, dP)x = [\int_{\Omega_\mathcal{B}} h_x \, dP_{\mathcal{B}[x]}]x = (T|_{\mathcal{B}[x]})x = Tx$ we also have that $T_x x = x$. It is immediate from (12) that $C_x T_x = T_x C_x = T_x$. Moreover,

(13) $$\|T_x\| = \| \int_{\Omega_\mathcal{B}} h_x \chi_{\delta_x} \, dP \| \leq 4\|\mathcal{B}\| \cdot |h_x \chi_{\delta_x}|_P \leq 4\|\mathcal{B}\| \cdot |h_x|_P$$

$$\leq 4\|\mathcal{B}\| \sup_{w \in \Omega_\mathcal{B}} |h_x(w)| \leq 4\|\mathcal{B}\| \cdot \|T\|.$$

Since $T_x \in \langle \mathcal{B} \rangle_u^-$ there is a *unique* function $f_x \in C(\Omega_\mathcal{B})$ such that $T_x = \Phi(f_x)$. Then $\Phi(f_x)x = T_x x = x$. Moreover, $(I - C_x)\Phi(f_x) = (I - C_x)T_x = 0$ as $C_x T_x = T_x$. However, we also have that

$$(I - C_x)\Phi(f_x) = \Phi(\chi_{\Omega_\mathcal{B} \setminus \delta_x})\Phi(f_x) = \Phi(\chi_{\Omega_\mathcal{B} \setminus \delta_x} f_x).$$

Since Φ is injective, it follows that $\chi_{\Omega_\mathcal{B} \setminus \delta_x} f_x = 0$. Hence, the function f_x has the properties

(14) $$f_x \in C(\Omega_\mathcal{B}) \text{ with both } f_x \chi_{\Omega_\mathcal{B} \setminus \delta_x} = 0 \text{ and } \Phi(f_x)x = Tx.$$

Moreover, if g is another function with the properties (14), then it follows from (10) that $g = f_x$ in $C(\Omega_B)$. Hence, the properties (14) characterize f_x.

The next claim is, for each $x \in X$, that

$$(15) \qquad\qquad \delta_{Bx} = \delta(B) \cap \delta_x, \qquad B \in \mathcal{B}.$$

To see this fix $x \in X$ and $B \in \mathcal{B}$. Let $z := Bx$ and note that $C_z Bz = C_z Bx = BC_x x = Bx = z$, so that $C_{Bx} \le C_x B$. However, if $C_{Bx} < C_x B$, then the *non-zero* projection $G \in \mathcal{B}$ defined by $G := (C_x B - C_{Bx})$ satisfies $G \le C_x B$ and $GC_{Bx} = 0$. Moreover, $GBx = GC_{Bx} Bx = 0$ and $G(I - B) = 0$ (since $G \le C_x B$ implies that $GC_x B = G$ and so $G(I - B) = GC_x B(I - B) = 0$). Thus

$$Gx = GBx + G(I - B)x = 0 + 0 = 0$$

and so $(I - G)x = x$. By (8) we see that $C_x \le (I - G)$, i.e. $G \le (I - C_x)$. Hence, $G \le C_x B$ and $G \le (I - C_x)$ and so

$$G \le (C_x B) \wedge (I - C_x) = C_x B(I - C_x) = 0,$$

that is, $G = 0$ which is a contradiction. So, we must have $C_{Bx} = C_x B$. That is, $\Phi(\chi_{\delta_{Bx}}) = \Phi(\chi_{\delta_x})\Phi(\chi_{\delta(B)}) = \Phi(\chi_{\delta_x \cap \delta(B)})$. Then the injectivity of Φ establishes (15).

For each $x \in X$, we now establish the identity

$$(16) \qquad\qquad f_{Bx} = f_x \chi_{\delta(B)}, \qquad B \in \mathcal{B}.$$

Let $g := f_x \chi_{\delta(B)}$. If $w \in \Omega_B \backslash \delta(B)$, then clearly $g(w) = 0$ (as $\chi_{\delta(B)}(w) = 0$) and if $w \in \Omega_B \backslash \delta_x$, then $f_x(w) = 0$ by (14) and so again $g(w) = 0$. Accordingly, $g\chi_{\Omega_B \backslash(\delta_x \cap \delta(B))} = 0$ and so by (15) we also have that $g\chi_{\Omega_B \backslash \delta_{Bx}} = 0$. Moreover, by (11) and (14) we see that

$$\Phi(g)Bx = \Phi(f_x \chi_{\delta(B)})Bx = \Phi(f_x)\Phi(\chi_{\delta(B)})Bx = \Phi(f_x)BBx = B\Phi(f_x)x = BTx = TBx.$$

So, with f_x replaced by g and x replaced by Bx in (14) we see from the uniqueness statement after (14) that $g = f_{Bx}$. This establishes (16).

The next step is to verify the following property:

(17)   For each $x, y \in X$ we have that f_x and f_y coincide on $\delta_x \cap \delta_y$. The first observation is to note that we may assume $\delta_x = \delta_y$. For, if $z := C_y x$ and $w := C_x y$, then putting $B = C_y$ into (15) yields $\delta_z = \delta_y \cap \delta_x$ and putting $B = C_x$ and replacing x by y in (15) yields $\delta_w = \delta_x \cap \delta_y$. Hence, $\delta_z = \delta_y \cap \delta_x = \delta_w$. Similar substitutions into (16) show that $f_z = f_x \chi_{\delta_y}$ and $f_w = f_y \chi_{\delta_x}$. Thus, if f_z and f_w coincide on $\delta_z = \delta_w$, then (17) will hold. So, we may assume that $\delta_x = \delta_y$.

Since $T(x - y) = Tx - Ty$ it follows from (14) that

$$\Phi(f_{x-y})(x - y) = T(x - y) = Tx - Ty = \Phi(f_x)x - \Phi(f_y)y$$

which, upon rearrangement, yields

$$(18) \qquad\qquad \Phi(f_{x-y} - f_x)x = \Phi(f_{x-y} - f_y)y.$$

To complete the proof of (17) we proceed by contradiction. So, suppose that $f_x \neq f_y$ on $\delta_x (= \delta_y)$. Choose a point $\lambda_0 \in \delta_x$ such that $f_x(\lambda_0) \neq f_y(\lambda_0)$. Then also $f_{x-y}(\lambda_0) - f_x(\lambda_0) \neq f_{x-y}(\lambda_0) - f_y(\lambda_0)$ and hence, at least one of $f_{x-y}(\lambda_0) - f_x(\lambda_0)$ or $f_{x-y}(\lambda_0) - f_y(\lambda_0)$ is non-zero. So, one of the functions f_x or f_y, say f_x, differs from both f_y and f_{x-y} at some point $\lambda_0 \in \delta_x$. Since $Co(\Omega_B)$ forms a base for the topology of Ω_B and $\{P(\delta(B)) : B \in \mathcal{B}\} = \mathcal{B}$, there exists some element $B \in \mathcal{B}$ and $\varepsilon > 0$ such that $\delta(B) \subseteq \delta_x = \delta_y$, with $\lambda_0 \in \delta(B)$, and

$$|f_{x-y}(w) - f_x(w)| \geq \varepsilon, \qquad w \in \delta(B).$$

Since $\delta(B) \in Co(\Omega_B)$, the function $g := \chi_{\delta(B)}/(f_{x-y} - f_x)$ belongs to $C(\Omega_B)$. Let $h := g(f_{x-y} - f_y)$. Then it follows from (18) that

$$\Phi(h)y = \Phi(g)\Phi(f_{x-y} - f_y)y = \Phi(g)\Phi(f_{x-y} - f_x)x = \Phi(g(f_{x-y} - f_x))x = \Phi(\chi_{\delta(B)})x = Bx,$$

and so by (14) (i.e. $\Phi(f_x)x = Tx$) and (11) (i.e. $\Phi(h)T = T\Phi(h)$) we have that

$$\begin{aligned}
\Phi(f_x)Bx &= B\Phi(f_x)x = BTx = TBx = T\Phi(h)y = \Phi(h)Ty \\
&= \Phi(h)\Phi(f_y)y = \Phi(f_y)\Phi(h)y = \Phi(f_y)Bx.
\end{aligned}$$

It follows that

$$\Phi(f_y\chi_{\delta(B)})Bx = \Phi(f_y)\Phi(\chi_{\delta(B)})Bx = \Phi(f_y)BBx = \Phi(f_y)Bx = TBx,$$

and that $f_y(w)\chi_{\delta(B)}(w) = 0$ for all $w \notin \delta_y \cap \delta(B) = \delta_x \cap \delta(B) = \delta_{Bx}$ (using (15)). Since the equations (14) determine f_x uniquely, we have that $f_y\chi_{\delta(B)} = f_{Bx}$. On the other hand, (16) shows that also $f_{Bx} = f_x\chi_{\delta(B)}$ and hence $f_x = f_y$ on $\delta(B)$, contradicting the fact that $\lambda_0 \in \delta(B)$ and $f_x(\lambda_0) \neq f_y(\lambda_0)$. This contradiction yields (17).

Because of (17) the function

$$\phi(w) := \left\{ \begin{array}{ll} f_x(w), & w \in \delta_x \\ 0, & w \notin \cup_{x \in X}\delta_x, \end{array} \right.$$

is well defined and continuous on the open set $\cup_{x \in X}\delta_x$. Since

$$\|f_x\|_\infty = \|\Phi^{-1}(T_x)\| \leq \|\Phi^{-1}\| \cdot \|T_x\| \leq 4\|\Phi^{-1}\| \cdot \|B\| \cdot \|T\|, \qquad x \in X,$$

(see (13)) we see that ϕ is a bounded Borel function. Define

(19)
$$T_0 := \int_\Omega \phi \, dP.$$

Then, for each $x \in X$, we have

$$\begin{aligned}
T_0x &= T_0C_xx = T_0P(\delta_x)x = \int_{\delta_x} \phi \, dPx = \int_{\delta_x} f_x \, dPx = P(\delta_x)\int_{\Omega_B} f_x \, dPx = \\
&= P(\delta_x)\Phi(f_x)x = \Phi(f_x)P(\delta_x)x = \Phi(f_x)C_xx = \Phi(f_x)x = Tx.
\end{aligned}$$

Accordingly, $T = T_0$ and it follows from (19) that $T_0 \in \langle P(Bo(\Omega_B))\rangle_u^- = \langle \mathcal{B}\rangle_u^-$. ∎

The following consequence is usually referred to as the *Bade reflexivity theorem*.

Corollary VII.3.1. *Let X be a Banach space and $B \subseteq L(X)$ be a Bade σ-complete B.a. of projections. Then*

$$\langle B \rangle_w^- = \langle B \rangle_s^- = \mathrm{AlgLat}(B).$$

Proof. Let $\overline{B}$ be the closure of B in $L_s(X)$. Then $\overline{B}$ is a Bade complete B.a. (c.f. Theorem V.8) and so, by Corollary V.6.3, we have that $\langle \overline{B} \rangle_w^- = \langle \overline{B} \rangle_s^- = \langle \overline{B} \rangle_u^-$. Combining this observation with $\langle \overline{B} \rangle_u^- = \mathrm{AlgLat}(\overline{B})$ (c.f. Theorem VII.3) and the identities $\langle \overline{B} \rangle_s^- = \langle B \rangle_s^-$ (c.f. Exercise 52(c)) and $\mathrm{AlgLat}(B) = \mathrm{AlgLat}(\overline{B})$ (which follows from Exercise 61(a)) yields the desired conclusion. ∎

Exercise 64. It was shown in the proof of Theorem VII.3 that $\mathrm{AlgLat}(B) \subseteq B^c$; an examination of that part of the proof shows this is the case for *any* B.a. of projections $B \subseteq L(X)$. Give an example to show that, in general, this inclusion is strict. ∎

Exercise 65. Let X be a Banach space and $B \subseteq L(X)$ be a Bade σ-complete B.a. of projections with the property that, for each $x \in X$, there exists a projection $P \in B^c$ such that $PX = B[x]$. Show that $\langle B \rangle_s^- = B^{cc}$. ∎

A single operator $T \in L(X)$ is called *reflexive* if the closed algebra in $L_s(X)$ generated by $\{I, T\}$ is reflexive. Of course, this closed algebra consists of the closure, in $L_s(X)$, of

$$\{p(T) : \ p \text{ a complex polynomial}\}.$$

D. Sarason showed in [39] that every bounded normal operator in a Hilbert space is reflexive. T.A. Gillespie extended this result to the Banach space setting by showing that every scalar-type spectral operator is a reflexive operator; see [21].

We end this chapter with an application of the Bade reflexivity theorem to deduce a classical result of J. von Neumann , usually referred to as the *bicommutant theorem*.

Theorem VII.4. *Let X be a separable Hilbert space and $T \in L(X)$ be a selfadjoint operator with resolution of the identity $P_T : Bo(\mathbb{R}) \longrightarrow L_s(X)$, so that $T = \int_{\mathbb{R}} \lambda \, dP_T(\lambda) = \int_{\sigma(T)} \lambda \, dP_T(\lambda)$. Then the following seven algebras are the same.*

(a) $\{\int_{\mathbb{R}} \phi \, dP_T : \ \phi : \mathbb{R} \longrightarrow \mathbb{C} \ a \ bounded \ Borel \ function\}$.

(b) *The closed subalgebra of $L_w(X)$ generated by $\{I, T\}$.*

(c) *The bicommutant $\{T\}^{cc}$, of T.*

(d) $\langle P_T(Bo(\mathbb{R})) \rangle_u^-$.

(e) $\langle P_T(Bo(\mathbb{R})) \rangle_s^- = \langle P_T(Bo(\mathbb{R})) \rangle_w^-$.

(f) $\mathrm{AlgLat}(P_T(Bo(\mathbb{R})))$.

(g) $\mathrm{AlgLat}(\{T\})$.

Proof. Let $\mathcal{A}_1, \ldots, \mathcal{A}_7$ denote the algebras defined by (a), ..., (g), respectively.

That $\mathcal{A}_2 = \mathcal{A}_7$ is precisely the result of D. Sarason referred to above.

Let $\mathcal{B} := P_T(Bo(\mathbb{R}))$. Then the separability of X implies that $\mathcal{B}$ is Bade complete (c.f. Proposition VI.3) and hence, $\mathcal{A}_4 = \mathcal{A}_5$ by Corollary V.6.3.

The equality $\mathcal{A}_4 = \mathcal{A}_6$ is precisely the Bade reflexivity theorem (see Corollary VII.3.1).

Theorem V.1 and an examination of the proof of Theorem V.2 show that $\mathcal{A}_1 = \mathcal{A}_4$.

Since $T = \int_{\mathbb{R}} \phi \, dP_T$, where $\phi(t) := t\chi_{\sigma(T)}(t)$, for $t \in \mathbb{R}$, is an element of $L^\infty(P_T) = L^1(P_T)$, and P_T is a *closed* spectral measure, it follows from Theorem V.6 that $T \in \mathcal{A}_5$. Since $\mathcal{A}_5$ is closed in $\mathcal{L}_w(X)$ and is a subalgebra, it follows that $\mathcal{A}_2 \subseteq \mathcal{A}_5$. To prove that $\mathcal{A}_5 \subseteq \mathcal{A}_2$ it suffices to show that $P_T(\delta) \in \mathcal{A}_2$, for all $\delta \in Bo(\mathbb{R})$. Since $\mathcal{A}_2$ is an algebra of operators, the family Σ of all Borel sets $\delta \subseteq \mathbb{R}$ for which $P_T(\delta) \in \mathcal{A}_2$ is an *algebra of sets* . Let $\{\delta_n\}_{n=1}^\infty \subseteq \Sigma$ be a monotone sequence with limit δ. By the countable additivity of P_T in the weak operator topology it follows that $P(\delta_n) \longrightarrow P(\delta)$ in $\mathcal{L}_w(X)$. Since $\{P(\delta_n)\}_{n=1}^\infty \subseteq \mathcal{A}_2$ and $\mathcal{A}_2$ is closed in $\mathcal{L}_w(X)$ we see that $P(\delta) \in \mathcal{A}_2$. Accordingly, $\delta \in \Sigma$. So, Σ is a σ-*algebra of sets* . Let $[a, b]$ be a bounded closed interval with $\sigma(T) \subseteq (a, b)$. Let $[u, v] \subseteq (a, b)$ and ϕ_ε be the continuous, piecewise affine function such that $\phi_\varepsilon = \mathbb{1}$ on $[u, v]$ and $\phi_\varepsilon = 0$ on $(-\infty, u - \varepsilon] \cup [v + \varepsilon, \infty)$, where $\varepsilon > 0$ satisfies $a < (u - \varepsilon)$ and $(v + \varepsilon) < b$. Then there exists a sequence of polynomials $\{q_n\}_{n=1}^\infty$ such that $q_n \longrightarrow \phi_\varepsilon$ in $C([a, b])$. By the dominated convergence theorem for P_T (see Corollary V.2.2), after noting that $\operatorname{supp}(P_T) = \sigma(T) \subseteq [a, b]$ and

$$|q_n|_{P_T} \leq \sup_{t \in [a,b]} |q_n(t)| \leq M, \qquad n \in \mathbb{N},$$

for some constant $M > 0$, we have that

$$q_n(T) = \int_{\sigma(T)} q_n \, dP_T = \int_{[a,b]} q_n \, dP_T \longrightarrow \int_{[a,b]} \phi_\varepsilon \, dP_T := \phi_\varepsilon(T), \qquad n \to \infty,$$

where the convergence is in $\mathcal{L}_s(X)$. Hence, $\phi_\varepsilon(T) \in \mathcal{A}_2$ for every such ε. Let $\varepsilon_n \downarrow 0$, in which case $\phi_{\varepsilon_n} \longrightarrow \chi_{[u,v]}$ pointwise on $[a, b]$, and use the dominated convergence theorem again yields

$$\lim_{n \to \infty} \phi_{\varepsilon_n}(T) = \lim_{n \to \infty} \int_{[a,b]} \phi_{\varepsilon_n} \, dP_T = \lim_{n \to \infty} \int_{[a,b]} \chi_{[u,v]} \, dP_T = P_T([u, v]),$$

where the convergence is again in $\mathcal{L}_s(X)$. Hence, $P_T([u, v]) \in \mathcal{A}_2$ and so $[u, v] \in \Sigma$. We note that $u = v$ is allowed and so also $\{u\} \in \Sigma$, for each $u \in (a, b)$. Since Σ is a σ-algebra it follows that Σ contains all subintervals of $[a, b]$. But, $P_T(\mathbb{R} \backslash [a, b]) = 0$ and so it follows, for an arbitrary interval $J \subseteq \mathbb{R}$, that

$$P_T(J) = P_T(J \backslash [a, b]) + P_T(J \cap [a, b]) = P_T(J \cap [a, b]) \in \mathcal{A}_2.$$

Hence, Σ contains all intervals from $\mathbb{R}$ and so $\Sigma = Bo(\mathbb{R})$. This establishes that $\mathcal{A}_2 = \mathcal{A}_5$.

It remains to show that $\mathcal{A}_2 = \mathcal{A}_3$. The inclusion $\mathcal{A}_2 \subseteq \mathcal{A}_3$ is clear. For the converse, let $Y \in \operatorname{Lat}(\mathcal{B})$ and Q be the orthogonal projection of X onto Y. Fix $x \in X$, in which case

$Qx \in Y$. If $B \in \mathcal{B}$, then $BY \subseteq Y$ and so $BQx \in Y$. Hence, $Q(BQx) = BQx$. Since x is arbitrary it follows that $QBQ = BQ$. Take adjoints of this identity gives $QBQ = QB$ (using $B^* = B$ and $Q^* = Q$). So, we see that $BQ = QB$ for all $B \in \mathcal{B}$, that is,

$$(20) \qquad P_T(\delta)Q = QP_T(\delta), \qquad \delta \in Bo(\mathbb{R}).$$

Since $T = \int_\mathbb{R} \lambda \, dP_T(\lambda)$ is the limit (in $\mathcal{L}_s(X)$) of operators of the form $\int_\mathbb{R} s \, dP_T$, where s is a Borel-simple function, it is clear from (20) that $TQ = QT$. Given $R \in \mathcal{A}_3$ it follows that $RQ = QR$. Since Q is the orthogonal projection onto Y it follows that $RY \subseteq Y$. So, we have shown that every operator from $\mathcal{A}_3$ belongs to AlgLat($\mathcal{B}$). That is, $\mathcal{A}_3 \subseteq \mathcal{A}_6$ ($= \mathcal{A}_2$ as already shown) and so $\mathcal{A}_3 \subseteq \mathcal{A}_2$. Hence, $\mathcal{A}_2 = \mathcal{A}_3$. ∎

If X is a *separable Banach space* and we use the fact that $\mathcal{A}_2 = \mathcal{A}_7$ (by the paper of T.A. Gillespie mentioned above in place of D. Sarason's paper), then an examination of the proof of Theorem VII.4 shows, for T a scalar-type spectral operator from $\mathcal{L}(X)$ with $\sigma(T) \subseteq \mathbb{R}$ and resolution of the identity P_T, that

$$(21) \qquad \mathcal{A}_1 = \mathcal{A}_2 = \mathcal{A}_4 = \mathcal{A}_5 = \mathcal{A}_6 = \mathcal{A}_7 \subseteq \mathcal{A}_3.$$

We remark that the last inclusion may be *strict*, even in a separable , reflexive Banach space. An example can be found in an elegant paper of J. Dieudonné, [9] . For some positive results, where the inclusion in (21) is actually an *equality*, we refer to [32]. In particular, if the range of P_T is *atomic* , then this is the case. Hence, for any *compact* scalar-type spectral operator T satisfying $\sigma(T) \subseteq \mathbb{R}$ the inclusion in (21) is an equality.

More generally, if X is an arbitrary Banach space and $\mathcal{B} \subseteq \mathcal{L}(X)$ is any Bade σ-complete B.a. of projections (not necessarily the resolution of the identity of some scalar-type spectral operator) one may ask when is it the case that

$$(22) \qquad \langle \mathcal{B} \rangle_s^- = \mathcal{B}^{cc} ?$$

It is shown in [32; Proposition 1] that (22) holds whenever $\mathcal{B}$ is *atomic* . This is of some interest since there exist classes of Banach spaces X in which *every* Bade σ-complete B.a. of projections is automatically atomic. For instance, if X is a Grothendieck space with the Dunford-Pettis property, or a hereditarily indecomposable space, or a space with the Schur property, or a complemented subspace of an $\mathcal{L}^\infty$-space, then this is the case; see [32; Proposition 2]. We note that there also exist other classes of Banach spaces X (besides Hilbert spaces) for which (22) holds for *all* Bade σ-complete B.a. 's of projections $\mathcal{B} \subseteq \mathcal{L}(X)$ and such that *non-atomic* B.a. 's $\mathcal{B}$ exist in $\mathcal{L}(X)$. For example, this is known to be the case if X is any complemented subspace of an $\mathcal{L}^1$-space, [30; Corollary 8], since any Bade complete B.a. of projections $\mathcal{B}$ in such a space X necessarily satisfies the hypotheses of Exercise 64; see [30; Theorem 7]. For the definition, examples and theory of $\mathcal{L}^p$-spaces we refer to [4] and the references there in. We conclude with the remark that (22) also holds in any Banach space X for any bounded B.a. of projections $\mathcal{B} \subseteq \mathcal{L}(X)$ which is τ-complete and has a cyclic vector, even in the stronger form

$$\langle \mathcal{B} \rangle_u^- = \langle \mathcal{B} \rangle_s^- = \mathcal{B}^{cc}.$$

This is an easy consequence of Theorem VII.2(b). As noted before, every Bade complete B.a. of projections is τ-complete.

Bibliography

[1] Bade, W.G., On Boolean algebras of projections and algebras of operators, *Trans. Amer. Math. Soc.* 80 (1955), 345-360.

[2] Bade, W.G., A multiplicity theory for Boolean algebras of projections in Banach spaces, *Trans. Amer. Math. Soc.* 92 (1959), 508-530.

[3] Bessaga, C. and Pełczynski, A., On bases and unconditional convergence of series in Banach spaces, *Studia Math.* 17 (1958), 151-164.

[4] Bourgain, J., *New Classes of $\mathcal{L}^p$-spaces*, Lecture Notes in Mathematics No. 889, Springer–Verlag, Berlin–Heidelberg–New York, 1981.

[5] Curbera, G.P., Operators into L^1 of a vector measure and applications to Banach lattices, *Math. Ann.* 292 (1992), 317-330.

[6] Curbera, G.P., When L^1 of a vector measure is an AL-space, *Pacific J. Math.* 162 (1994), 287-303.

[7] Curbera, G.P., Banach space properties of L^1 of a vector measure, *Proc. Amer. Math. Soc.* 123 (1995), 3797-3806.

[8] Diestel, J. and Uhl, J.J. Jr., *Vector Measures*, Amer. Math. Soc., Providence, 1977.

[9] Dieudonné, J., Sur la bicommutante d'une algèbre d'opérateurs, *Portugaliae Math.* 14 (1955), 35-38.

[10] Dixmier, J., *Von Neumann Algebras*, North Holland, Amsterdam, 1981.

[11] Dodds, P.G. and Ricker, W.J., Spectral measures and the Bade reflexivity theorem, *J. Funct. Anal.* 61 (1985), 136-163.

[12] Dodds, P.G., de Pagter, B. and Ricker, W.J., Reflexivity and order properties of scalar-type spectral operators in locally convex spaces, *Trans. Amer. Math. Soc.* 293 (1986), 355-380.

[13] Dowson, H.R., *Spectral Theory of Linear Operators*, Academic Press, London, 1978.

[14] Dunford, N. and Schwartz, J.T., *Linear Operators I: General Theory*, Wiley–Interscience Publ. (2nd Printing), New York, 1964.

[15] Dunford, N. and Schwartz, J.T. *Linear Operators III: Spectral Operators*, Wiley–Interscience Publ., New York, 1971.

[16] Dwinger, Ph., *Introduction to Boolean Algebras*, Physica–Verlag, Würzburg, 1961.

[17] Engelking, R., *General Topology*, PWN–Polish Scientific Publ., Wrocław, 1977.

[18] Fell, J.M.G. and Kelly, J.L., An algebra of unbounded operators, *Proc. Nat. Acad. Sci. U.S.A.*, 38 (1952), 592-598.

[19] Fernández, A., Naranjo, F. and Ricker, W.J., Completeness of L^1-spaces for measures with values in complex vector spaces, *J. Math. Anal. Appl.*, 223 (1998), 76-87.

[20] Gardner, R.J., The regularity of Borel measures and Borel measure-compactness, *Proc. London Math. Soc. (3)*, 30 (1975), 95-113.

[21] Gillespie, T.A., Boolean algebras of projections and reflexive algebras of operators, *Proc. London Math. Soc.* 37 (1978), 56-74.

[22] Gillespie, T.A., Strongly closed bounded Boolean algebras of projections, *Glasgow Math. J.* 22 (1981), 73-75.

[23] Gillespie, T.A., Bade functionals, *Proc. Roy. Irish Acad. Sect. A*, 81A (1981), 13-23.

[24] Gillman, L. and Jerison, M., *Rings of Continuous Functions*, D. van Nostrand Co., New York, 1960.

[25] Halmos, P.R., Commutativity and spectral properties of normal operators, *Acta Sci. Math. (Szeged)*, 12 (1950), 153-156.

[26] Halmos, P.R., *Lectures on Boolean Algebras*, van Nostrand Reinhold Co., London, 1963.

[27] Kluvánek, I. and Knowles, G., *Vector Measures and Control Systems*, North Holland, Amsterdam, 1975.

[28] Koppelberg, S., *Handbook of Boolean Algebras*, Vol. 1, North Holland, Amsterdam, 1989.

[29] Lewis, D.R., Integration with respect to vector measures, *Pacific J. Math.* 33 (1970), 157-165.

[30] McCarthy, C.A., and Tzafriri, L., Projections in $\mathcal{L}_1$ and $\mathcal{L}_\infty$-spaces, *Pacific J. Math.* 26 (1968), 529-546.

[31] de Pagter, B. and Ricker, W.J., Boolean algebras of projections and resolutions of the identity of scalar-type spectral operators, *Proc. Edinburgh Math. Soc.* 40 (1997), 425-435.

[32] de Pagter, B. and Ricker, W.J., On the bicommutant theorem for σ-complete Boolean algebras of projections in Banach spaces, *Indag. Math. (New Series)*, 10 (1999), 87-100.

[33] Rall, C. Über Boolesche Algebren von Projektionen, *Math. Z.* 153 (1977), 199-217.

[34] Ricker, W.J., Criteria for closedness of vector measures, *Proc. Amer. Math. Soc.* 91 (1984), 75-80.

[35] Ricker, W.J., Separability of the L^1-space of a vector measure, *Glasgow Math. J.* 34 (1992), 1-9.

[36] Ricker, W.J., Rybakov's theorem in Fréchet spaces and completeness of L^1-spaces, *J. Austral. Math. Soc. (Series A)*, 64 (1998), 247-252.

[37] Royden, H.L., *Real Analysis (2nd Edition)*, Macmillan Co. New York, 1968.

[38] Rudin, W., *Real and Complex Analysis (2nd Edition)*, McGraw–Hill Book Co., New York–St. Louis–San Francisco, 1974.

[39] Sarason, D., Invariant subspaces and unstarrred operator algebras, *Pacific J. Math.* 17 (1966), 511-517.

[40] Sikorski, R., *Boolean Algebras (2nd Edition)*, Springer–Verlag, Berlin–Göttingen–Heidelberg, 1964.

[41] Walker, R.C., *The Stone-Čech Compactification*, Springer–Verlag, Berlin–Heidelberg, 1974.

Appendix

The following collection of references is (hopefully) a comprehensive list of all papers on the general topic of spectral operators and Boolean algebras of projections which have appeared since 1979 or are not recorded in the monograph *Linear Operators III: Spectral Operators* (by N. Dunford and J.T. Schwartz, Wiley-Interscience, New York, 1971) or in the monograph *Spectral Theory of Linear Operators* (by H.R. Dowson, Academic Press, London, 1978). Any omissions are unintentional.

Abramovich, Y.A., Arenson, E.L., and Kitover, A.K.

1. Operators in Banach $C(K)$-modules and their spectral properties. (Russian) *Dokl. Akad. Nauk SSSR* **301** (1988), no.3, 525-528; *translation in Soviet Math. Dokl.* **38** (1989), 93-97. MR90a:47091.

2. Banach $C(K)$-modules and operators preserving disjointness, Pitman Research Notes in Mathematics, 277, *Longman Scientific and Technical, Harlow*, 1992. MR94d:47027.

Akhmedov, A.M.

1. Perturbations of compact spectral operators. (Russian) *Izv. Akad. Nauk Azerbaĭdzhan. SSR Ser. Fiz.-Tekhn. Mat. Nauk* **5** (1984), 22-27. MR86e:47041.

2. A perturbation of compact spectral operators. (Russian) *Akad. Nauk Azerbaĭdzhan. SSR Dok.* **40** (1984), 10-13. MR86g:47043.

3. Spectrality of perturbations of compact operators. (Russian) *Linear operators and their applications (Russian)*, 3-14, Azerbaĭdzhan. Gos. Univ., Baku, 1986. MR90j:47039.

Albrecht, E.

1. On some classes of generalized spectral operators. *Arch. Math. (Basel)* **30** (1978), 297-303. MR57#10486.

2. A characterization of spectral operators on Hilbert spaces. *Glasgow Math. J.* **23** (1982), 91-95. MR83a:47040.

Albrecht, E., and Ricker, W.J.

1. Local spectral properties of constant coefficient differential operators in $L^p(\mathbb{R}^N)$. *J. Operator Theory* **24** (1990), 85-103. MR92b:47048.

2. Local spectral properties of certain matrix differential operators in $L^p(\mathbb{R}^N)^m$. *J. Operator Theory* **35** (1996), 3-37. MR98b:47060.

Al-Khezi, S.

1. Analytic functions of a prespectral operator. *Glasgow Math. J.* **23** (1982), 171-175. MR83h:47022.

Al-Khezi, S., and Dowson, H.R.

1. Quasispectral operators, *Proc. Roy. Irish Acad. Sect. A* **81** (1981), 25-28. MR82m:47026.

Amrein, W. (see Jauch, J.M.)

Anderson, J., and Foiaş, C.

1. Properties which normal operators share with normal derivations and related operators. *Pacific J. Math.* **61** (1975), 313-325. MR54#1010.

Andruchow, E., Recht, L., and Stojanoff, D.

1. The space of spectral measures is a homogeneous reductive space. *Integral Equations Operator Theory* **16** (1993), 1-14. MR93j:46078.

Applebaum, D.

1. Spectral families of quantum stochastic integrals. *Comm. Math. Phys.* **170** (1995), 607-628. MR96f:81061.

Apostol, C.

1. Invariant subspaces for subquasiscalar operators, *J. Operator Theory.* **3** (1980), 159-164. MR83a:47005a.

2. The spectral flavour of Scott Brown's techniques, *J. Operator Theory.* **6** (1981), 3- 12. MR83a:47005b.

Arenson, E.L. (see Abramovich, Y.A.)

Azoff, E.A.

1. A note on direct integrals of spectral operations. *Michigan Math. J.* **23** (1976), 65-69. MR55#13280.

Azzouni, A.

1. Opérateurs spectraux à points critiques. (French) [Spectral operators with critical points] *Bull. Math. Soc. Sci. Math. Roumanie (N.S.)* **35-(83)** (1991), 3-14. MR95k:47051.

Bacalu, I.

1. Residual spectral measures. (Romanian) *Stud. Cerc. Math.* **27** (1975), 377-379. MR53#3777.

2. Restrictions and quotients of spectral systems. (Romanian) *Stud. Cerc. Mat.* **32** (1980), 113-118. MR81h:47031.

Bai, F.D. (see Li, Y.S.)

Bartle, R.G.

1. Selfadjoint operators and some generalizations. *Operator theory and functional analysis (Papers, Summer Meeting, Amer. Math. Soc., Providence, R.I., 1978)*, pp36-50, Res. Notes in Math., 38, *Pitman, Boston, Mass.-London*, 1979. MR81f:47029.

Baskakov, A.G.

1. Methods of abstract harmonic analysis in the theory of perturbations of linear operators (Russian), *Sibirsk. Mat. Zh.* **24** (1983), no.1, 21-39, 191. MR85j:47010.

2. The method of similar operators and formulas for regularized traces. (Russian), *Izv. Vyssh. Uchebn. Zaved. Mat.* **1984**, no.3, 3-12. MR86a:47009.

3. Regularized trace formulas for powers of perturbed spectral operators. (Russian), *Izv. Vyssh. Uchebn. Zaved. Mat.* **1985**, 68-71, 86. MR87i:47017.

4. Spectral analysis with respect to finite-dimensional perturbations of spectral operators. (Russian), *Izv. Vyssh. Uchebn. Zaved. Mat.* **1991**, 3-11. MR92m:47023.

5. Spectral analysis of perturbed non-quasi-analytic and spectral operators. (Russian) *Izv. Ross. Akad. Nauk Ser. Mat.* **58** (1994), 3-32; *translation in Russian Acad. Sci. Izv. Math.* **45** (1995), 1-31. MR96d:47004.

Bauer, G., and Mennicken, R.

1. Störungstheorie für diskrete Spektraloperatoren und Anwendungen auf im Parameter nichtlineare Eigenwertprobleme. (German) [Perturbation theory for discrete spectral operators and applications to eigenvalue problems that are nonlinear in the parameter] Regensburger Mathematische Schriften [Regensburg Mathematical Publications], **10**. *Universität Regensburg, Fachbereich Mathematik, Regensburg*, 1985. vii+104pp. MR87k:47028.

Berezanski, Yu. M.

1. The projection spectral theorem, *Uspekhi Mat. Nauk* **39**:4(1984), 3-52; *Russian Math. Surveys* **39**:4(1984), 1-62. MR86e:47029.

2. Projection spectral theorem and its applications to the infinite-dimensional harmonic analysis. *Gaussian random fields (Nagoya, 1990)*, 114-128, Ser. Probab. Statist., 1, *World Sci. Publishing, River Edge, NJ*, 1991. MR93b:47032.

Berezanski, Yu. M., Zhernakov, N.W., and Us, G.F.

1. A spectral approach to quantum stochastic integrals. *Rep. Math. Phys.* **28** (1989), 347-360. MR93:47142.

Bernstein, A.R.

1. The spectral theorem–a nonstandard approach. *Z. Math. Logik Grundlagen Math.* **18** (1972), 419-434. MR47#4048.

Birman, M. Sh., and Solomyak, M.Z.

1. Operator integration, perturbations and commutators. (Russian) *Zap. Nauchn. Sem. Leningrad. Otdel. Mat. Inst. Steklov. (LOMI)* **170** (1989), Issled. Linein. Oper. Teorii Funktsii. 17, 34-66, 321. MR91b:47086.

2. Tensor product of a finite number of spectral measures is always a spectral measure. *Integral Equations Operator Theory* **24** (1996), 179-187. MR96m:47038.

Birman, M. Sh., Vershik, A.M., and Solomyak, M.Z.

1. The product of commuting spectral measures may fail to be countably additive. (Russian) *Funktsional. Anal. i Prilozhen.* **13** (1979), 61-62. MR80h:28010.

Blashchak, V.A.

1. On a differential operator of the second order on the whole axis with spectral singularities, *Dopovidi Akad. Nauk Ukrain. RSR* (1966), 38-41 (Ukrainian).

Burnap, C., and Zweifel, P.F.

1. A note on the spectral theorem. *Integral Equations Operator Theory* **9** (1986), 305-324. MR87h:47078.

Byrne, C.M.

1. Banach function spaces and spectral measures, *PhD Thesis, University of Edinburgh,* 1982.

Byrne, C.M., and Gillespie, T.A.

1. The representation of spectral measures. *Proc. Roy. Irish Acad. Sect. A* **85** (1985), 31-42. MR87e:47040.

Cao, X.D.

1. Some results on spectral systems of commuting operators. (Chinese) *Acta Sci. Natur. Univ. Jilin.* **1993**, 27-33. MR96c:47045.

Chabauty, R.

1. Opérateurs semi-simples. (French) [Semisimple operators] *Nonlinear analysis,* 1982/1983, Exp. No.10, 40pp., Publ. Math. Fac. Sci. Besançon, 7, *Univ. Franche-Comté, Besançon,* 1983. MR86h:47025.

Cheng, Qingpeng

1. Well-bounded operators on general Banach spaces. *PhD Thesis, Murdoch University*, 1999.

Chourasia, N.N.

1. Decomposable and spectral operators. *Indian J. Pure Appl. Math.* **12** (1981), 73- 80. MR82g:47025.

2. On Weyl's theorem for spectral operators and essential spectra of direct sum. *Pure Appl. Math. Sci.* **15** (1982), 39-45. MR84b:47039.

Cleaver, C.E.

1. A characterization of spectral operators of finite type. *Compositio Math.* **26** (1973), 95-99. MR47#9336.

Conway, J.B., and Gillespie, T.A.

1. Is a selfadjoint operator determined by its invariant subspace lattice? *J. Funct. Anal.* **64** (1985), 178-189. MR87h:47041.

Curtain, R.F.

1. Spectral systems. *Internat. J. Control* **39** (1984), 657-666. MR85h:93039.

Čuǐko, G.I. (see Ljance, V.É.)

Davidson, K.R.

1. Essentially spectral operators. *Proc. London Math. Soc. (3)* **46** (1983), 547-560. MR84i:47048.

Dayanithy, K.

1. Interpolation of spectral operators. *Math. Z.* **159** (1978), 1-2. MR80a:47048.

Delanghe, R., and Van hamme, J.

1. Generalised spectral measures. *Proc. Roy. Irish Acad. Sect. A* **87** (1987), 17-26. MR89b:47050.

Dinescu, G.

1. On semiscalar operators. *Rev. Roumaine Math. Pures Appl.* **28** (1983), 359-380. MR85k:47055.

Dodds, P.G.

1. Boolean algebras of projections in locally convex spaces. *Miniconference on operator theory and partial differential equations (Canberra, 1983)*, 67-76, Proc. Centre Math. Anal. Austral. Nat. Univ., **5**, *Austral. Nat. Univ., Canberra*, 1984. MR86a:47044.

2. Scalar type spectral opeators in locally convex spaces. *Miniconference on linear analysis and function spaces (Canberra, 1984)*, 185-193, Proc. Centre Math. Anal. Austral. Nat. Univ., **9**, *Austral. Nat. Univ., Canberra*, 1985. MR825524.

Dodds, P.G., and de Pagter, B.

1. Orthomorphisms and Boolean algebras of projections, *Math. Z.* **187** (1984), 361- 381. MR86a:47045.

2. Algebras of unbounded scalar-type spectral operators. *Pacific J. Math.* **130** (1987), 41-74. MR89e:47070.

Dodds, P.G., and Ricker, W.J.

1. Spectral measures and the Bade reflexivity theorem. *J. Funct. Anal.* **61** (1985), 136-163. MR86i:47042.

Dodds, P.G., de Pagter, B., and Ricker, W.J.

1. Reflexivity and order properties of scalar-type spectral operators in locally convex spaces. *Trans. Amer. Math. Soc.* **293** (1986), 355-380. MR87d:47046.

Doust, I.

1. Contractive projections on Banach spaces. *Miniconference on Fucntional Analysis & Optimization*, Proc. Centre Math. Anal., ANU, **20** (1988), 50-58. MR90i:46020.

2. Well-bounded and scalar-type spectral operators on spaces not containing c_0. *Proc. Amer. Math. Soc.* **105** (1989), 367-370. MR89f:47048.

3. Well-bounded and scalar-type spectral operators on L^p-spaces. *J. London Math. Soc. (2)* **39** (1989), 525-534. MR90f:47043.

4. An example in the theory of spectral and well-bounded operators. *Miniconference on Operators in Analysis (Sydney, 1989)*, 83-90, Proc. Centre Math. Anal. Austral. Nat. Univ., **24**, *Austral. Nat. Univ., Canberra*, 1990. MR91k:47071.

5. Interpolation and extrapolation of well-bounded operators. *J. Operator Theory* **28** (1992), 229-250. MR95i:47064.

6. A weaker condition for normality. *Glasgow Math. J.* **36** (1994), 249-253. MR95f:47041.

Doust, I., and de Laubenfels, R.

1. Functional calculus, integral representations, and Banach space geometry. *Quaestiones Math.* **17** (1994), 161-171. MR95e:47028.

Doust, I., and Ricker, W.J.

1. Spectral properties for Hermitian operators. *Linear Algebra Appl.* **175** (1992), 75- 96. MR93g:47024.

Dowson, H.R. (see also Al-Khezi, S.)

1. Spectral theory of linear operators. London Mathematical Society Monographs, 12. *Academic Press, Inc. [Harcourt Brace Jovanovich, Publishers], London-New York*, 1978. xii+422 pp.ISBN:0-12-220950-8. MR80c:47022.

2. Restrictions of scalar-type spectral operators. *Bull. London Math. Soc.* **10** (1978), 305-309. MR80a:47050.

3. Solution to a problem of Lior Tzafriri. *Bull. London Math. Soc.* **23** (1991), 285- 292. MR92j:47083.

Dowson, H.R., and Gillespie, T.A.

1. A representation theorem for a complete Boolean algebra of projections. *Proc. Roy. Soc. Edinburgh Sect. A* **83** (1979), 225-237. MR80m:47039.

Drewnowski, L., Florencio, M., and Paúl, P.

1. The space of Pettis integrable functions is barrelled. *Proc. Amer. Math. Soc.* **114** (1992), 687-694. MR92f:46045.

2. Uniform boundedness of operators and barrelledness in spaces with Boolean algebras of projections, *Atti. Sem. Mat. Fis. Univ. Modena* **41** (1993), 317-329. MR94m:46004.

3. Barrelled subspaces of spaces with subseries decompositions or Boolean rings of projections, *Glasgow Math. J.* **36** (1994), 57-69. MR94m:46005.

Dubrovskiĭ, V.V. and Sadovničiĭ, V.A.

1. Unbounded perturbations of spectral operators. (Russian) *Problems in mathematical physics and numerical mathematics (Russian)*, pp. 137-144, 325. *Nauka, Moscow*, 1977. MR58#30463.

Duncan, R.

1. Weak convergence of spectral mesaures, *Ann. Univ. Mariae Curie-Sklodowska Sect. A* **51** (1997), 35-39. MR99a:47004.

Dudley, R.M.

1. A note on products of spectral measures. *Vector and operator valued measures and applications (Proc. Sympos., Alta, Utah, 1972)*, pp. 125-126. *Academic Press, New York*, 1973. MR49#1193.

Dunford, N.

1. An expansion theorem. *Proceedings of the Conference on Integration, Topology, and Geometry in Linear Spaces (Univ. North Carolina, Chapel Hill, N.C., 1979)*, pp. 61-73, Contemp. Math., **2**, *Amer. Math. Soc., Providence, R.I.*, 1980. MR82i:47048.

2. Spectral theory in topological vector spaces. *Functions, series, operators, Vol.I, II (Budapest, 1980)*, 391-422, Colloq. Math. Soc. János Bolyai, **35**, *North-Holland, Amsterdam-New York*, 1983. MR86a:47029.

Emamirad, H. (see de Laubenfels, R.)

Erdos, J.A.

  1. On Boolean algebras of projections. *Glasgow Math. J.* **18** (1977), 69-72. MR55#1131.

Farwig, R., and Marschall, E.

  1. On the type of spectral operators and the nonspectrality of several differential operators on L_p. *Integral Equations Operator Theory* **4** (1981), 206-214. MR82b:47037.

Faulkner, G.D., and Huneycutt, J.E., Jr.

  1. The canonical form of a scalar operator on a Banach space. *Proc. Amer. Math. Soc.* **71** (1978), 81-84. MR58#7196.

Fialkow, L.

  1. A note on quasisimilarity. II. *Pacific J. Math.* **70** (1977), 151-162. MR57#17341.

Fixman, U., and Tzafriri, L.

  1. The full algebra generated by a spectral operator. *J. London Math. Soc. (2)* **4** (1971), 39-45. MR45#943.

Fleming, R.J., and Jamison, J.E.

  1. Classes of operators on vector valued integration spaces. *J. Austral. Math. Soc. Ser. A* **24** (1977), 129-138. MR58#2403.

Florencio, M. (see Drewnowski, L.)

Foaiş, C. (see Anderson, J.)

Folland, G.B.

  1. Spectral analysis of a singular non-selfadjoint boundary value problem. *J. Differential Equations.* **37** (1980), 206-224. MR81m:34022.

  2. Spectral analysis of a non-selfadjoint differential operator, *J. Differential Equations.* **39** (1981), 151-185. MR82m:34021.

Fong, C.K., and Lam, L.

  1. On spectral theory and convexity. *Trans. Amer. Math. Soc.* **264** (1981), 59-75. MR82c:46061.

Fong, C.K., and Radjabalipour, M.

  1. On quasi-affine transforms of spectral operators. *Michigan Math. J.* **23** (1976), 147-150. MR53#14207.

Förster, K.-H., and Nagy, B.

1. The spectral singularities of injective weighted shifts. *Acta Math. Hungar.* **56** (1990), 121-124. MR92b:47037.

Fox, D.W.

1. Spectral measures and separation of variables. *J. Res. Nat. Bur. Standards Sect. B* **80B** (1976), 347-351. MR54#13615.

Frunză, S.

1. Some results on duality for spectral decompositions, Spectral Theory: Banach Center Publ. No.8, pp.263-279. *PWN-Polish Scientific Publ., Warsaw*, 1982. MR85m:47032.

Gaudry, G.I., and Ricker, W.J.

1. Spectral properties of L_p translations. *J. Operator Theory* **14** (1985), 87- 111. MR86i:47043.

2. Spectral properties of translation operators in certain function spaces. *Illinois J. Math.* **31** (1987), 453-468. MR89f:47047.

Georgescu, V. (see Jauch, J.M.)

Gillespie, T.A. (see also Byrne, C.M., Conway, J.B., and Dowson, H.R.)

1. Cyclic Banach spaces and reflexive operator algebras. *Proc. Roy. Soc. Edinburgh Sect. A* **78** (1977/78), 225-235. MR57#13566.

2. Boolean algebras of projections and reflexive algebras of operators. *Proc. London Math. Soc. (3)* **37** (1978), 56-74. MR58#2433.

3. Strongly closed bounded Boolean algebras of projections. *Glasgow Math. J.* **22** (1981), 73-75. MR82a:46018.

4. Spectral measures on spaces not containing ℓ_∞. *Proc. Edinburgh Math. Soc. (2)* **24** (1981), 41-45. MR82g:47026.

5. Bade functionals. *Proc. Roy. Irish Acad. Sect. A* **81** (1981), 13-23. MR83c:47052.

6. Boundedness criteria for Boolean algebras of projections. *J. Funct. Anal.* **148** (1997), 70-85.

Ginzburg, Ju. P.

1. Spectral measures and the duality of spectral subspaces of contractions with a slowly growing resolvent. (Russian) Investigations on linear operators and the theory of functions, VIII. *Zap. Nauchn. Sem. Leningrad. Otdel. Mat. Inst. Steklov. (LOMI)* **73** (1977), 203-206. MR80a:47025.

Goldstein, R.A. (see Saeks, R.)

Grandis, M.

1. Cohesive categories and measurable operators. *Rend. Accad. Naz. Sci. XL Mem. Mat. (5)* **14** (1990), 195-234. MR93b:46140.

Hadwin, D.

1. Fuglede's theorem and limits of spectral operators. *Proc. Amer. Math. Soc.* **68** (1978), 365-368. MR58#12372.

Hadwin, D., and Orhon, M.

1. Reflexivity and approximate reflexivity for bounded Boolean algebras of projections. *J. Funct. Anal.* **87** (1989), 348-358. MR91e:47047.

2. A noncommutative theory of Bade functionals. *Glasgow Math. J.* **33** (1991), 73- 81. MR91m:46085.

Harvey, B.N.

1. Spectral operators with critical points. *Amer. J. Math.* **96** (1974), 41-61. MR51#3949.

Helson, H.

1. The spectral theorem. Lecture Notes in Math. No.1227, *Springer-Verlag, Berlin-New York*, 1986. vi + 104pp. ISBN: 3-540-17197-5. MR88k:47027.

Hladnik, M.

1. On prespectrality of generalised derivations. *Proc. Roy. Soc. Edinburgh Sect. A* **104** (1986), 93-106. MR88b:47042.

2. When are generalized derivations spectral? *Operators in indefinite metric spaces, scattering theory and other topics (Bucharest, 1985)*, 215-226, Oper. Theory: Adv. Appl., **24**, *Birkhäuser, Basel-Boston, MA*, 1987. MR89a:47050.

3. Spectrality of elementary operators. *J. Austral. Math. Soc. Ser. A* **49** (1990), 327-346. MR91k:47077.

Holland, W.F.

1. Eigenstructure specification for linear systems in Hilbert space. *PhD Thesis, Macquarie University*, 1987.

Hou, X.Z., and Zou, C.Z.

1. The spectral decompositions of reducible operators on Banach spaces. (Chinese) *Chinese Ann. Math. Ser. A* **9** (1988), 598-603. MR90g:47061.

Huang, S.Z.

1. Each hyperinvariant subspace for a multiplication operator is spectral. *Proc. Amer. Math. Soc.* **106** (1989), 1057-1061. MR90h:47017.

Hughes, R.J. (see Kantorovitz, Sh.)

Huige, G.E.

1. The spectral theory of some non-selfadjoint differential operators. *Comm. Pure Appl. Math.* **21** (1968), 25-49. MR39#3352.

2. Perturbation theory of some spectral operators. *Comm. Pure Appl. Math.* **24** (1971), 741-757. MR46#694.

3. The spectral resolution of some non-selfadjoint partial differential operators. *Canad. J. Math.* **27** (1975), 1316-1322. MR53#1050.

Huneycutt, J.E., Jr. (see Faulkner, G.D.)

Hyb, W.

1. On the spectral properties of translation operators in one-dimensional tubes, *Ann. Polon. Math.* **55** (1991), 157-161. MR92k:47060.

Jamison, J.E. (see Fleming, R.J.)

Janas, J.

1. Semi-spectral integrals and related mappings. *Studia Math.* **46** (1973), 301-313. MR49#1194.

2. The spectral theorem in its historical aspect. (Polish) *Opuscula Math.* No.13 (1993), 7, 14, 37-43. MR94m:47002.

Jauch, J.M., Amrein, W., and Georgescu, V.

1. The spectral integral in scattering theory. In: *Vector and operator-valued measures and applications (Proc. Sympos. Snowbird, Alta, Utah, 1972)*, pp.149-153, *Academic Press, New York*, 1973. MR48#10374.

Jazar, M. (see de Laubenfels, R.)

Jean, M.D.

1. On the adjoint of spectral operators. *Bull. Inst. Math. Acad. Sinica* **11** (1983), 257-260. MR85k:47054.

Jefferies, B.R.F., and Ricker, W.J.

1. On the additivity of C.A. McCarthy's product spectral measure. *Integral Equations Operator Theory* **15** (1992), 706-708. MR93d:47066.

Jibril, A.A.S.

1. On complete Boolean algebra of projections. *Arabian J. Sci. Engrg.* **9** (1984), 51- 53. MR85e:47051.

Junggeburth, J., and Nessel, R.J.

1. Multipliers with respect to Abel-bounded spectral measures in locally convex spaces. *Rev. Un. Mat. Argentina* **28** (1976/77), 46-53. MR58#6930.

Kalb, K.G.

1. The spectral measure of spectral operators of finite type. (Spanish) *Rev. Colombiana Mat.* **13** (1979), 265-270. MR82d:47043.

Kantorovitz, Sh. (see also de Laubenfels, R.)

1. Spectral equivalence and Volterra elements. *Indiana Univ. Math. J.* **22** (1972/73), 951-957. MR53#8957.

2. Spectrality criteria for unbounded operators with real spectrum. *Math. Ann.* **256** (1981), 19-28. MR82i:47049.

3. Characterization of unbounded spectral operators with spectrum in a half-line. *Comment. Math. Helv.* **56** (1981), 163-178. MR83a:47042.

4. Spectral theory of Banach space operators. C^k-classification, abstract Volterra operators, similarity, spectrality, local spectral analysis. Lecture Notes in Mathematics, 1012. *Springer- Verlag, Berlin-New York*, 1983. ii+179 pp. ISBN: 3-540-12673-2. MR85g:47001.

5. Semigroups of operators and spectral theory. Pitman Research Notes in Mathematics Series, 330. *Longman Scientific & Technical, Harlow; copublished in the United States with John Wiley & Sons, Inc., New York*, 1995. x+135 pp. ISBN:0-582-27778-7. MR97c:47045.

Kantorovitz, Sh., and Hughes, R.J.

1. Spectral representation of local semigroups. *Math. Ann.* **259** (1982), 455- 470. MR83i:47051.

2. Spectral representations for unbounded operators with real spectrum. *Math. Ann.* **282** (1988), 535-544. MR90a:47009.

Kitover, A.K. (see Abramovich, Y.A.)

Kluvánek, I.

1. Applications of vector measures. *Proceedings of the Conference on Integration, Topology, and Geometry in Linear Spaces (Univ. North Carolina, Chapel Hill, N.C., 1979)*, pp.101-134, Contemp. Math., **2**, *Amer. Math. Soc., Providence, R.I.*, 1980. MR82m:28019.

2. Operator valued measures and perturbations of semigroups, *Arch. Rational Mech. Anal.* **81** (1983), 161-180. MR84j:28019.

137

3. Some problems of spectral theory. *Miniconference on operator theory and partial differential equations (Canberra, 1983)*, 103-108, Proc. Centre Math. Anal. Austral. Nat. Univ., **5**, *Austral. Nat. Univ., Canberra*, 1984.

4. Scalar operators, *Linear Algebra Appl.* **84** (1986), 399-402.

5. Integration for the spectral theory. *Miniconference on operator theory and partial differential equations (North Ryde, 1986)*, 26-34, Proc. Centre Math. Anal. Austral. Nat. Univ., **14**, *Austral. Nat. Univ., Canberra*, 1986. MR88k:47044.

6. Scalar operators and integration. *J. Austral. Math. Soc. Ser. A* **45** (1988), 401-420. MR89j:47023.

7. Banach algebras occurring in spectral theory. *Conference on Automatic Continuity and Banach Algebras (Canberra, 1989)*, 239-253, Proc. Centre Math. Anal. Austral. Nat. Univ., **21**, *Austral. Nat. Univ., Canberra*, 1989. MR90i:47036.

Kluvánek, I., and Ricker, W.J.

1. Spectral set functions and scalar operators. *Integral Equations Operator Theory* **17** (1993), 277-299. MR94i:47049.

Kocan, D.

1. A characterization of some spectral manifolds for a class of operators. *Illinois J. Math.* **16** (1972), 359-369. MR46#2474.

Kristóf, J.

1. Commutative GW^*-algebras, *Acta Sci. Math. (Szeged)*. **52** (1988), 145- 155. MR90c:46067.

2. Spectral theorem for normal elements of GW^*-algebras. *Studia Sci. Math. Hungar.* **23** (1988), 453-469. MR90f:46112.

Labrousse, J.-P.

1. Opérateurs spectraux et opérateurs quasi normaux. (French) *C.R. Acad. Sci. Paris Sér. A-B* **286** (1978), A1107-A1108. MR58#12496.

Lam, L. (see Fong, C.K.)

Lange, R., and Nagy, B.

1. Semigroups and scalar-type operators in Banach spaces. *J. Funct. Anal.* **119** (1994), 468-480. MR95h:47054.

de Laubenfels, R. (see also Doust, I.)

1. The moment problem and C^n-scalar operators. *Honam Math. J.* **7** (1985), 7-13. MR87c:47023.

  2. Unbounded scalar operators on Banach lattices. *Honam Math. J.* **8** (1986), 1-19. MR88e:47064.

  3. Scalar-type spectral operators and holomorphic semigroups. *Semigroup Forum* **33** (1986), 257-263. MR88a:47031.

  4. Spectral projections, semigroups of operators, and the Laplace transform. *Linear operators (Warsaw, 1994)*, 193-204, Banach Center Publ., **38**, *Polish Acad. Sci., Warsaw*, 1997. MR98d:47042.

de Laubenfels, R., and Kantorovitz, Sh.

  1. The semi-simplicity manifold on aribtrary Banach spaces. *J. Funct. Anal.* **133** (1995), 138-167. MR96g:47012.

de Laubenfels, R., Emamirad, H., and Jazar, M.

  1. Regularized scalar operators. *Appl. Math. Lett.* **10** (1997), 65-69. MR97m:47044.

Lennon, M.J.J.

  1. Direct integral decomposition of spectral operators, *Math. Ann.* **207** (1974), 257- 268. MR49#5922.

Li, B.R.

  1. Perturbation theory for discrete spectral operators. (Chinese) *Kexue Tongbao* **1978**, 282-285. MR82f:47022.

  2. Perturbation theory for a class of linear operators and its applications (Chinese. English summary). *Acta Math. Sinica* **21** (1978), 206-222. MR80c:47018.

  3. The perturbation theory for linear operators of discrete type. *Pacific J. Math.* **104** (1983), 29-38. MR84a:47039.

Li, J.H. (see Sun, S.L.)

Li, Y.R.

  1. Perturbation and logarithms of some prespectral operators. *Nanjing Daxue Xuebao Shuxue Bannian Kan.* **11** (1994), 152-159. MR96g:47025.

Li, Y.S., Bai, F.D., and Zou, C.Z.

  1. Spaces of *(OP)* type and spectral operators. (Chinese) *J. Harbin Inst. Tech.* **24** (1992), 10-13. MR94d:47034.

Liu, J. (see Luo, Y.H.)

Liu, L.F.

  1. On completely spectral operators. (Chinese) *Chinese Ann. Math.* **2** (1981), 61-64. MR82f:47047.

Ljance, V.È., and Čuĭko, G.I.

1. Operator measures. (Russian) *Dokl. Akad. Nauk Ukrain. SSR Ser. A* **1981**, 7-9. MR82k:47034.

Locker, J.

1. The nonspectral Birkhoff-regular differential operators determined by $-D^2$. *J. Math. Anal. Appl.* **154** (1991), 243-254. MR91m:47069.

Luo, Y.H.

1. Root properties of a class of discrete spectral operators. (Chinese) *Acta Math. Sinica* **32** (1989), 556-563. MR91c:47064.

Luo, Y.H., and Liu, J.

1. On the growth order of C_0-semigroups I. *Appl. Math. J. Chinese Univ. Ser. B.* **10** (1995), 69-74. MR96c:47055.

Lutz, D.

1. Scalar spectral operators, ordered ℓ^p-direct sums, and the counterexample of Kakutani-McCarthy. *Pacific J. Math.* **62** (1976), 497-505. MR54#3484.

2. A perturbation theorem for spectral operators. *Pacific J. Math.* **68** (1977), 127-134. MR56#6460.

3. Über operatorwertige Lösungen der Funktionalgleichung des Cosinus. (German) *Math. Z.* **171** (1980), 233-245. MR81i:47043.

Maczy'nski, M.J.

1. Extension properties for spectral measures. *Bull. Acad. Polon. Sci. Sér. Sci. Math. Astronom. Phys.* **26** (1978), 35-39. MR57#10487.

2. A Banach space reformulation of the spectral theorem. *Demonstratio Math.* **15** (1982), 851-860. MR84e:47025.

Marschall, E. (see Farwig, R.)

Masani, P., and Rosenberg, M.

1. When is an operator the integral of a given spectral measure? *J. Funct. Anal.* **21** (1976), 88-121. MR53#6347.

Maserick, P.H.

1. Spectral theory of operator-valued transformations. *J. Math. Anal. Appl.* **41** (1973), 497-507. MR49#7828.

Mbekhta, M.

1. Généralisation de la décomposition de Kato aux opérateurs paranormaux et spectraux. (French) [Generalization of the Kato decomposition to paranormal and spectral operators] *Glasgow Math. J.* **29** (1987), 159-175. MR88i:47010.

McEnnis, B.W.

1. Shifts on Krein spaces. Operator theory: operator algebras and applications, Part 2 (Durham, NH 1988), 201-211. *Proc. Sympos. Pure Math*, **51**, Part 2. *Amer. Math. Soc., Providence*, 1990. MR91j:47038.

McIntosh, A., Pryde, A., and Ricker, W.J.

1. Comparison of joint spectra for certain classes of commuting operators. *Studia Math.* **88** (1988), 23-36. MR89e:47006.

Mennicken, R. (see Bauer, G.)

Mockenhaupt, G., and Ricker, W.J.

1. Idempotent multipliers for $L^p(\mathbb{R})$. *Arch. Math. (Basel)*, to appear.

Mukherjee, R.N.

1. On operators quasisimilar to spectral operators. *Tamkang J. Math.* **9** (1978), 99- 101. MR80g:47039.

Muraz, G.

1. Opérateurs subordonnés à une mesure spectrale. (French) *C.R. Acad. Sci. Paris Sér. A-B* **289** (1979), A271-A273. MR80i:47036.

2. Opérateurs non bornés subordonnés à une mesure spectrale. (French) *C.R. Acad. Sci. Paris Sér. A-B* **289** (1979), A341-A343. MR81a:47027.

Nagy, B. (see also Förster, K.-H., and Lange, R.)

1. Analytic functions of prespectral operators. *Acta Math. Acad. Sci. Hungar.* **31** (1978), 157-171. MR58#23737.

2. Unbounded prespectral operators. *Period. Math. Hungar.* **9** (1978), 277-283. MR80a:47052.

3. Characterizations of spectral operators. *Arch. Math. (Basel)* **32** (1979), 289-294. MR80m:47031.

4. Essential spectra of spectral operators. *Period. Math. Hungar.* **11** (1980), 1-6. MR81e:47031.

5. Residually spectral operators. *Acta Math. Acad. Sci. Hungar.* **35** (1980), 37-48. MR81k:47045.

6. On Boolean algebras of projections and prespectral operators. *Invariant subspaces and other topics (Timişoara/Herculane, 1981)*, pp.145-162, Operator Theory: Adv. Appl., **6**, *Birkhäuser, Basel-Boston, Mass.*, 1982. MR84f:47038.

7. On Boolean algebras of projections and prespectral operators. International conference on analytical methods in number theory and analysis (Moscow, 1981). *Trudy Mat. Inst. Steklov.* **163** (1984), 187-190. MR86b:47056.

8. Operators with spectral singularities. *J. Operator Theory* **15** (1986), 307-325. MR87f:47047.

9. Finitely spectral operators. *Glasgow Math. J.* **28** (1986), 95-112. MR87k:47077.

10. Spectral measures with singularities. *Acta Math. Hungar.* **49** (1987), 51-64. MR87k:47078.

11. On Stone's theorem in Banach space. *Acta Sci. Math. (Szeged)* **57** (1993), 207- 213. MR94j:47058.

12. On KLKH systems of scalar-type spectral operators. *Evolution equations (Baton Rouge, LA, 1992)*,317-326, Lecture Notes in Pure and Appl. Math., 168, *Dekker, New York*, 1995. MR95j:47042.

Napiórkowski, K.

1. On spectral representations of Banach spaces. *Bull. Acad. Polon. Sci. Sér. Sci. Math. Astonom. Phys.* **20** (1972), 841-844. MR47#2416.

Nessel, R.J. (see Junggeburth, J.)

Niechwiej, J.

1. Support of the joint spectral measure. *J. Austral. Math. Soc. Ser. A* **43** (1987), 74-80. MR88d:47030.

Nukherjee, R.N.

1. A note on normality of operators. *Tamkang J. Math.* **11** (1980), 33-35. MR82j:47027.

Okada, S.

1. Spectra of scalar-type spectral operators and Schauder decompositions. *Math. Nachr.* **139** (1988), 167-174. MR89m:47028.

Okada, S., and Ricker, W.J.

1. Uniform operator σ-additivity of indefinite integrals induced by scalar-type spectral operators. *Proc. Roy. Soc. Edinburgh Sect. A* **101** (1985), 141-146. MR87f:47048.

2. Nonweak compactness of the integration map for vector measures. *J. Austral. Math. Soc. Ser. A* **54** (1993), 287-303. MR94d:28014.

3. Compactness properties of vector-valued integration maps in locally convex spaces. *Colloq. Math.* **67** (1994), 1-14. MR95j:28009.

4. Spectral measures which fail to be equicontinuous. *Period. Math. Hungar.* **28** (1994), 55-61. MR95k:47052.

5. Criteria for weak compactness of vector-valued integration maps. *Comment. Math. Univ. Carolin.* **35** (1994), 485-495. MR95m:28009.

6. Continuous extensions of spectral measures. *Colloq. Math.* **71** (1996), 115-132. MR98b:47033.

7. Spectral measures and automatic continuity. *Bull. Belg. Math. Soc. Simon Stevin* **3** (1996), 267-279. MR98b:47034.

8. Boolean algebras of projections and ranges of spectral measures, *Dissertationes Math.* **365**, 33p., 1997.

9. Representation of complete Boolean algebras of projections as ranges of spectral measures. *Acta Sci. Math. (Szeged),* **63** (1997), 209-227; see also Errata in same journal, **63** (1997) 689-693.

10. Integration with respect to the canonical spectral measure in sequence spaces. *Collect. Math.*, to appear.

11. Criteria for closedness of spectral measures and completeness of Boolean algebras of projections. *J. Math. Anal. Appl.*, **232** (1999), 197-221.

Omladič, M.

1. On *n*-spectral operators. *Rev. Roumaine Math. Pures Appl.* **32** (1987), 451-464. MR88k:47045.

Orhon, M. (see also Hadwin, D.)

1. Boolean algebras of commuting projections. *Math. Z.* **183** (1983), 531-537. MR85a:47049.

2. Algebras of operators containing a Boolean algebra of projections of finite multiplicity. *Operators in indefinite metric spaces, scattering theory and other topics (Bucharest, 1985),* 265- 281, Oper. Theory: Adv. Appl., **24**, *Birkhäuser, Basel-Boston, Mass.,* 1987. MR88k:47063.

3. Boolean algebras of projections on Banach spaces. National Mathematics Symposium (Trabzon, 1987). *J. Karadeniz Tech. Univ. Fac. Arts Sci. Ser. Math.-Phys.* **11** (1988), 21-32 (1989). MR92b:47050.

Owusu-Ansah, T.

1. Hermitian operators, symmetric operators and the spectral theorem on certain Banach spaces. *Afrika Mat.* **4** (1982), 5-23. MR83b:47036.

de Pagter, B. (see also Dodds, P.G.)

1. A note on a paper by W.J. Ricker and H.H. Schaefer: "The uniformly closed algebra generated by a complete Boolean algebra of projections". *Math. Z.* **203** (1990), 645-647. MR91i:47050.

de Pagter, B., and Ricker, W.J.

1. Boolean algebras of projections and resolutions of the identity of scalar-type spectral operators. *Proc. Edinburgh Math. Soc.* **40** (1997), 425-435.

2. On the bicommutant theorem for σ-complete Boolean algebras of projections in Banach spaces. *Indag. Math. (N.S.).* **10** (1999), 87-100.

Paliogiannis, F.C.

1. Topological function-theoretic proofs in spectral theory. *Rend. Circ. Mat. Palermo (2)* **44** (1995), 21-44. MR96e:47027.

Palled, S.V. (see Panchapagesan, T.V.)

Panchapagesan, T.V.

1. Extension of spectral measures. *Illinois J. Math.* **16** (1972), 130-142. MR46#2473.

2. A note on spectral and prespectral operators. *Rev. Colombiana Mat.* **16** (1982), 57-63. MR84a:47041.

3. Some characterizations of extendible spectral measures. (Spanish) *Rev. Colombiana Mat.* **21** (1987), 65-72. MR90f:47044.

4. Unitary invariants of spectral measures. *Proceedings of the Ramanujan Centennial International Conference (Annamalainagar, 1987)*, 103-117, RMS Publ., 1, *Ramanujan Math. Soc., Annamalainagar*, 1988. MR90h:47042.

5. Uniform ordered spectral decompositions. *Rev. Colombiana Mat.* **24** (1990), 41- 49. MR92b:46118.

6. Unitary invariants of spectral measures with the CGS-property. *Rend. Circ. Mat. Palermo (2)* **42** (1993), 219-248. MR94k:47034.

7. On simple spectral measures. *Volume in homage to Dr. Rodolfo A. Ricabarra (Spanish)*, 83-97, Vol. Homenaje, 1, *Univ. Nac. del Sur, Bahía Blanca*, 1995. MR96i:47040.

8. Orthogonal and bounded orthogonal spectral representations. *Rend. Circ. Mat. Palermo (2)* **44** (1995), 417-440. MR4388755.

Panchapagesan, T.V., and Palled, S.V.

1. On vector lattice-valued measures I, *Math. Slovaca*, **33** (1983), 269-292. MR85e:28013.

Patel, A.B.

1. A joint spectral theorem for unbounded normal operators. *J. Austral. Math. Soc. Ser. A* **34** (1983), 203-213. MR84i:47039.

Paúl, P. (see Drewnowski, L.)

Plafker, S.

1. The spectral theorem in Banach algebras. *Glasgow Math. J.* **13** (1972), 49- 55. MR46#7952.

Pryde, A. (see McIntosh, A.)

Räbiger, F., and Ricker, W.J.

1. C_0-semigroups and cosine families of linear operators in hereditarily indecomposable Banach spaces. *Acta Sci. Math. (Szeged)*, **64** (1998), 697-706.

2. Aspects of operator theory in hereditarily indecomposable Banach spaces. *Rend. Sem. Mat. Fis. Milano,* to appear.

Radjabalipour, M. (see also Fong, C.K.)

1. Growth conditions, spectral operators and reductive operators. *Indiana Univ. Math. J.* **23** (1973/74), 981-990. MR49#3591.

2. Local inverses of spectral operators. *Spectral theory of linear operators and related topics (Timişoara/Herculane, 1983),* 229-233, Operator Theory: Adv. Appl, **14**, *Birkhäuser, Basel-Boston, Mass.,* 1984. MR86i:47044.

3. A characterization of spectral operators. *Proc. Amer. Math. Soc.* **113** (1991), 167-170. MR91k:47073.

Radjavi, H., and Rosenthal, P.

1. Hyperinvariant subspaces for spectral and n-normal operators. *Acta Sci. Math. (Szeged)* **32** (1971), 121-126. MR46#7931.

Rall, C.

1. Boolesche Algebren von Projektionen auf Banachräumen, *PhD Thesis, Universität Tübingen,* 1975 (German).

2. Über Boolesche Algebren von Projektionen, *Math. Z.* **153** (1977), 199-217 (German). MR55#13270.

Recht, L. (see Andruchow, E.)

Ricker, W.J. (see also Albrecht, E., Dodds, P.G., Doust, I., Gaudry, G.I., Jefferies, B.R.F., Kluvánek, I., McIntosh, A., Mockenhaupt, G., Okada, S., de Pagter, B., and Räbiger, F.)

1. The product of spectral measures. *J. Operator Theory* **6** (1981), 351-361. MR83f:47018.

2. A criterion for an operator with real spectrum to be of scalar-type. *Monatsh. Math.* **95** (1983), 229-234. MR85a:47038.

3. On Boolean algebras of projections and scalar-type spectral operators. *Proc. Amer. Math. Soc.* **87** (1983), 73-77. MR84b:47040.

4. Extended spectral operators. *J. Operator Theory* **9** (1983), 269-296. MR84i:47049.

5. Closed spectral measures in Fréchet spaces. *Internat. J. Math. Math. Sci.* **7** (1984), 15-21. MR85m:46040.

6. Some problems of spectral theory. *Miniconference on operator theory and partial differential equations (Canberra, 1983)*, 145-152, Proc. Centre Math. Anal. Austral. Nat. Univ., **5**, *Austral. Nat. Univ., Canberra*, 1984. MR86b:47058.

7. Semigroups of operators and an application to spectral theory. *Math. Proc. Cambridge Philos. Soc.* **96** (1984), 143-149. MR86b:47057.

8. Spectral representation of local semigroups associated with Klein-Landau systems. *Math. Proc. Cambridge Philos. Soc.* **95** (1984), 93-100. MR85f:47045.

9. Characterization of Stieltjes transforms of vector measures and an application to spectral theory. *Hokkaido Math. J.* **13** (1984), 299-309. MR86h:47050.

10. Spectral respresentation of local semigroups in locally convex spaces. *Czechoslovak Math. J.* **35(110)** (1985), 248-259. MR86h:47066.

11. A spectral mapping theorem for scalar-type spectral operators in locally convex spaces. *Integral Equations Operator Theory* **8** (1985), 276-293. MR86j:47054.

12. Spectral operators of scalar type in Grothendieck spaces with the Dunford-Pettis property. *Bull. London Math. Soc.* **17** (1985), 268-270. MR87e:47041.

13. Stone's theorem for strongly continuous groups of isometries in Hardy spaces. *J. Funct. Anal.* **67** (1986), 206-227. MR87g:47088.

14. Spectral operators and weakly compact homomorphisms in a class of Banach spaces. *Glasgow Math. J.* **28** (1986), 215-222. MR87i:47047.

15. Countable additivity of multiplicative, operator-valued set functions. *Acta Math. Hungar.* **47** (1986), 121-126. MR87k:46097.

16. Well-bounded operators of type (B) in a class of Banach spaces. *J. Austral. Math. Soc. Ser. A* **42** (1987), 399-408. MR88b:47044.

17. Spectral measures, boundedly σ-complete Boolean algebras and applications to operator theory. *Trans. Amer. Math. Soc.* **304** (1987), 819-838. MR89k:47068.

18. Uniformly closed algebras generated by Boolean algebras of projections in locally convex spaces. *Canad. J. Math.* **39** (1987), 1123-1146. MR89k:47069.

19. A concrete realization of the dual space of L^1-spaces of certain vector and operator-valued measures. *J. Austral. Math. Soc. Ser. A* **42** (1987), 265-279. MR88m:46037.

20. Functional calculi for the Laplace operator in $L^p(\mathbb{R})$. *Miniconference on harmonic analysis and operator algebras (Canberra, 1987)*, 242-254, Proc. Centre Math. Anal. Austral. Nat. Univ., **15**, *Austral. Nat. Univ., Canberra*, 1987. MR90b:47086.

21. Spectral projections and the commutant of the Laplace operator in $L^p(\mathbb{R})$. *Rend. Sem. Mat. Fis. Milano* **57** (1987), 519-528 (1989). MR90j:47034.

22. "Spectral subsets" of $\mathbb{R}^m$ associated with commuting families of linear operators. *Proceedings of the analysis conference, Singapore 1986*, 243-247, North-Holland Math. Stud., 150, *North-Holland, Amsterdam-New York*, 1988. MR89e:47007.

23. Vector measures with values in the compact operators, *Proc. Amer. Math. Soc.* **102** (1988), 441-442. MR90f:28007.

24. Spectral properties of the Laplace operator in $L^p(\mathbb{R})$, *Osaka J. Math.* **25** (1988), 399-410. MR89h:47071.

25. Bade's theorem on the uniformly closed algebra generated by a Boolean algebra, *Comment. Math. Univ. Carolin.* **29** (1988), 597-600. MR90a:47112.

26. An L^1-type functional calculus for the Laplace operator in $L^p(\mathbb{R})$. *J. Operator Theory* **21** (1989), 41-67. MR91b:47109.

27. Operator algebras generated by Boolean algebras of projections in Montel spaces. *Inegral Equations Operator Theory*, **12** (1989), 143-145. MR90a:47113.

28. Analogues of von Neumann's bicommutant theorem. *Semesterbericht Funktionalanalysis Tübingen*, Wintersemester, **15** (1989), 213-221.

29. Spectral measures and integration: counterexamples. *Semesterbericht Funktionalanalysis Tübingen*, Sommersemester, **16** (1989), 123-129.

30. Boolean algebras of projections of uniform multiplicity one. *Miniconference on Operators in Analysis (Sydney, 1989)*, 206-212, Proc. Centre Math. Anal. Austral. Nat. Univ., **24** *Austral. Nat. Univ., Canberra*, 1990. MR91d:47031.

31. Boolean algebras of projections and spectral measures in dual spaces. *Linear operators in function spaces (Timişoara, 1988)*, 289-300, Oper. Theory Adv. Appl., **43**, *Birkhäuser, Basel*, 1990. MR92e:47059.

32. Scalar-type spectral operators and $C(\Omega)$-operational calculi. *Integral Equations Operator Theory* **13** (1990), 901-905. MR92k:47069.

33. Non-spectrality of generators of some classical analytic semigroups. *Indag. Math. (N.S.)* **1** (1990), 95-103. MR91b:47089.

34. Kakutani's example on product spectral measures. *Hokkaido Math. J.* **20** (1991), 593-600. MR92h:47049.

35. Spectral-like multipliers in $L^p(\mathbb{R})$. *Arch. Math. (Basel)* **57** (1991), 395- 401. MR92j:42009.

36. Integral transforms of vector measures on semigroups with applications to spectral operators. *Semigroup Forum* **45** (1992), 342-363. MR93h:28017.

37. Weak compactness of the integration map associated with a spectral measure. *Indag. Math. (N.S.)* **5** (1994), 353-364. MR95j:47026.

38. Well-bounded operators of type (B) in H.I. Spaces. *Acta Sci. Math. (Szeged)* **59** (1994), 475-488. MR96j:47027.

39. Weak compactness of integration maps associated with indefinite integrals of spectral measures. *Indag. Math. (N.S.)* **6** (1995), 495-503. MR97e:47047.

40. Spectrality for matrices of Fourier multiplier operators acting in L^p-spaces over lca groups. *Quaestiones Math.* **19** (1996), 237-257. MR97e:47048.

41. Existence of Bade functionals for complete Boolean algebras of projections in Fréchet spaces. *Proc. Amer. Math. Soc.* **125** (1997), 2401-2407. MR97j:47050.

42. The sequential closedness of σ-complete Boolean algebras of projections. *J. Math. Anal. Appl.* **208** (1997), 364-371. MR97m:47022.

43. The strong closure of σ-complete Boolean algebras of projections, *Arch. Math. (Basel)*, **72** (1999), 282-288.

44. Lipschitz continuity of spectral measures, *Bull. Austral. Math. Soc.*, **59** (1999), 369-373.

45. The spectral theorem: a historical viewpoint. *Ulmer Seminare: Funktionalanalysis und Differentialgleichungen.* To appear in vol.4, 1999.

Ricker, W.J., and Schaefer, H.H.

1. The uniformly closed algebra generated by a complete Boolean algebra of projections. *Math. Z.* **201** (1989), 429-439. MR91b:47081.

Rosenberg, M. (see Masani, P.)

Rosenthal, P., and Sourour, A.R. (see also Radjavi, H.)

1. On operator algebras containing cyclic Boolean algebras. *Pacific J. Math.* **70** (1977), 243-252. MR58#17876a.

2. On operator algebras containing cyclic Boolean algebras. II. *J. London Math. Soc.* (2) **16** (1977), 501-506. MR58#17876b.

Rybkin, A.V.

1. The Stieltjes B-integral and the summation of spectral decompositions of some classes of contraction operators. (Russian) *Dokl. Akad. Nauk SSSR* **319** (1991), 562-566; *translation in Soviet Math. Dokl.* **44** (1992), 166-170. MR92k:47024.

Sadovničiĭ, V.A. (see Dubrovskiĭ, V.V.)

Saeks, R., and Goldstein, R.A.

1. Cauchy integrals and spectral measures. *Indiana Univ. Math. J.* **22** (1972/73), 367-378. MR47#435.

Schaefer, H.H. (see also Ricker, W.J.)

1. Banach lattices and positive operators, *Springer-Verlag, New York-Heidelberg-Berlin*, 1974, xi +376pp. ISBN:0-387-06936-4. MR54#11023.

Sen, D.K.

1. Some results on Hermitian operators on Banach spaces. *Bull. Calcutta Math. Soc.* **77** (1985), 287-290. MR87i:47029.

Schonbek, T.P.

1. On a calculus for a generalized scalar operator. *J. Math. Anal. Appl.* **58** (1977), 527-540. MR58#7155.

Shobov, M.A.

1. Spectral operators generated by damped hyperbolic equations. *Integral Equations Operator Theory* **28** (1997), 358-372.

Shuchat, A.

1. Vector measures and the spectral theorem. In: *Vector and operator valued measures and applications (Proc. Sympos., Alta, Utah, 1972)*, pp.339-341. *Academic Press, New York*, 1973. MR49#3589.

2. Vector measures and scalar operators in locally convex spaces. *Michigan Math. J.* **24** (1977), 303-310. MR58#12497.

3. Spectral measures and homomorphisms. *Rev. Roumaine Math. Pures Appl.* **23** (1978), 939-945. MR80a:47053.

4. Infrabarreled spaces, linear operators and vector measures. *Bull. Soc. Roy. Sci. Liège,* **48** (1979), 153-157. MR81d:46002.

Smith, W.V.

1. The Kluvánek–Kantorovitz characterization of scalar operators in locally convex spaces. *Internat. J. Math. Math. Sci.* **5** (1982), 345-349. MR83g:47038.

2. Differential operators in Banach spaces with densely defined spectral measures, *Houston J. Math.* **8** (1982), 429-448. MR84d:34019.

3. Spectral measures. I. General theory in a topological vector space. *Period. Math. Hungar.* **13** (1982), 205-227. MR84d:46063.

4. Spectral measures. II. Characterization of scalar operators. *Period. Math. Hungar.* **13** (1982), 273-287. MR84g:47031.

5. Spectral measures. III. Densely defined spectral measures. *Period. Math. Hungar.* **15** (1984), 189-203. MR86d:47039.

Solomyak, M.Z. (see Birman, M. Sh.)

Sourour, A.R. (see also Rosenthal, P.)

1. On groups and semigroups of spectral operators on a Banach space. *Acta Sci. Math. (Szeged)* **36** (1974), 291-294. MR55#6239.

2. Semigroups of scalar type operators on Banach spaces, *Trans. Amer. Math. Soc.* **200** (1974), 207-232. MR51#1481.

3. Unbounded operators generated by a given spectral measure. *J. Funct. Anal.* **29** (1978), 16-22. MR81h:47020.

Spain, P.G.

1. Boolean algebras of projections. *Proc. Edinburgh Math. Soc. (2)* **19** (1974/75), 287-289. MR52#6497.

Stochel, J.

1. The Bochner-Kolmogorov extension theorem for semispectral measures. *Colloq. Math.* **54** (1987), 83-94. MR89b:47053.

Stojanoff, D. (see Andruchow, E.)

Stroescu, E.

1. Dunford subspectral systems of operators on a Banach space. (Romanian) *Stud. Cerc. Mat.* **35** (1983), 399-409. MR86a:47030.

2. $\mathcal{A}$-subspectral systems of operators on a Banach space. (Romanian) *Stud. Cerc. Mat.* **35** (1983), 518-528. MR86a:47031.

3. A Stone theorem for subscalar operators on Banach space. *Rev. Roumaine Math. Pures Appl.* **30** (1985), 571-578. MR87f:47061.

Šul'man, V.S.

1. Linear operator equations with generalized scalar coefficients. (Russian) *Dokl. Akad. Nauk SSSR* **225** (1975), 56-58=(English translation) *Soviet Math. Dokl.* **16** (1975), 1465-1468. MR58#7156.

Sun, S.L.

1. Decomposability for a class of nonspectral second-order ordinary differential operators. (Chinese) *Acta Sci. Natur. Univ. Jilin.* **1982**, no.4, 9-16. MR85b:34032.

Sun, S.L., and Li, J.H.

1. Similarity transformations and scalar operators. (Chinese) *Acta Sci. Natur. Univ. Jilin.* **1995**, no.3, 9-11. MR98d:47071.

Sussmann, H.J.

1. Non-spectrality of a class of second order ordinary differential operators, *Comm. Pure Appl. Math.* **23** (1970), 819-840. MR41#9055.

2. Generalized spectral theory and second order ordinary differential operators, *Canad. J. Math.* **25** (1973), 178-193.

Swartz, C.

1. An abstract uniform boundedness result. *Math. Slovaca,* **49** (1999), 63-69.

Syroid, I.-P.P.

1. A nonselfadjoint analogue of nonreflecting potentials. (Russian) *Mathematical analysis and probability theory (Russian),* pp.168-172, 222, "*Naukova Dumka*", Kiev, 1978. MR80m:34022.

2. Sufficient conditions for the absence of spectral singularities in a nonselfadjoint Schrödinger operator in terms of potential (Russian). *Sibirsk. Mat. Zh.* **22** (1981), 151-157. MR82f:34028.

3. Nonselfadjoint perturbation of the continuous spectrum of the Dirac operator (Russian). *Ukrain. Mat. Zh.* **35** (1983), 115-119. MR85c:34033.

4. Spectrality of the Dirac operator in terms of a potential. *Dokl. Akad. Nauk Ukr. SSR, Ser. A.* 1986, no.12, 8-10. MR88c:34034.

5. Conditions for the absence of spectral singularities in the non-selfadjoint Dirac operator in terms of the potential, *Ukrain. Mat. Zh.* **38** (1986), 352-359. MR88c:34035.

6. On the spectral property of the Dirac operator in terms of the potential, In: Methods for the investigations of differential and integral operators, *Collect. Sci. Works, Kiev.* 181-186, 1989. MR93e:47062.

Tanahashi, K.

1. Reductive weak decomposable operators are spectral. *Proc. Amer. Math. Soc.* **87** (1983), 44-46. MR84d:47043.

Tanahashi, K., and Yoshino, T.

1. A characterization of spectral operators on Hilbert spaces. *Proc. Amer. Math. Soc.* **90** (1984), 567-570. MR85h:47040.

2. Weak decomposable and spectral operators on a weakly complete Banach space. *Math. Japon.* **34** (1989), 831-835. MR91i:47051.

151

Tzafriri, L. (see Fixman, U.)

Us, G.F. (see Berezanski, Yu. M.)

Van hamme, J. (see also Delanghe, R.)

1. On operators generated by generalized spectral measures. *Simon Stevin* **62** (1988), 13-27. MR89i:47061.

Vasilescu, F.-H.

1. Analytic functional calculus and spectral decompositions, *D. Reidel Publishing Co., Dordrecht-Boston-London,* 1982. xiv + 378pp. ISBN:90-277-1376-6. MR85b:47016.

Vershik, A.M. (see Birman, M.Sh.)

Vieten, P.

1. Four characterizations of scalar-type operators with spectrum in a half-line. *Studia Math.* **122** (1997), 39-54. MR97m:47045.

Vu Kuok Fong (also spelt as Quôc-Phóng, Vũ)

1. On the spectral theory of scalar operators in Banach spaces. (Russian) *Dokl. Akad. Nauk SSSR* **254** (1980), 1038-1042. MR82c:47039.

2. On the theory of spectral operators of scalar type in Banach spaces. (Russian) *Math. Nachr.* **121** (1985), 319-344. MR87g:47063.

Wadhwa, B.L.

1. Decomposable and spectral operators on a Hilbert space. *Proc. Amer. Math. Soc.* **40** (1973), 112-114. MR48#943.

Wang, S.W.

1. $D_{\langle M_k \rangle}$ operators and spectral operators. *Chinese Ann. Math.* **1** (1980), 325-334. MR82j:47055.

Weibang, G.

1. Scalar-type operators that are commutators of compact operators. *An. Univ. Bucureşti Mat.* **32** (1983), 89-90. MR84m:47045.

Wu, P.Y.

1. Conditions for completely nonunitary contractions to be spectral. *J. Funct. Anal.* **31** (1979), 1-12. MR82d:47014.

Xu, F., and Zou, C.Z.

1. On the Boolean algebra of projection operators. (Chinese) *Dongbei Shida Xuebao* **1982**, no.1, 17-22. MR84g:47032.

Yakovlev, V.S.

1. Measurability of funtions with respect to spaces with resolutions of the identity. (Russian) *Spectral theory of operators and infinite-dimensional analysis*, 138-143, vi, *Akad. Nauk Ukrain. SSR, Inst. Mat., Kiev*, 1984. MR87a:47036.

Yoshino, T. (see Tanahashi, K.)

Yu, D.H.

1. The perturbation and spectrum distribution of a discrete spectral operator. (Chinese) *Sichuan Daxue Xuebao* **1982**, no.4, 11-19. MR84i:47050.

Zhernakov, N.W. (see Berezanski, Yu. M.)

Zou, C.Z. (see Hou, X.Z., Li, Y.S., and Xu, F.)

Zwart, H.

1. Characterization of all controlled invariant subspaces for spectral systems. *SIAM J. Control Optim.* **26** (1988), 369-386. MR89e:93129.

Zweifel, P.F. (see Burnap, C.)

List of symbols

The page reference indicates the page in the text where the symbol is first introduced **with an explanation**.

$\mathcal{B}$	27	$\vee B$	26
$\mathcal{E}$	26	B_Y	87
I	26	$\mathcal{B}_Y$	87
$\mathbb{C}$	1	c_0	3
$\mathbb{N}$	1	C_x	97
$\mathbb{R}$	1	d_x	59
$\emptyset$	1	δ_n	10
τ	31	χ_E	2
$\mathbb{1}$	29	E^c	1
$\ll$	12	f^+	16
$\leq$	25	f^-	16
$\subseteq$	1	I_m	21
$\overline{\mathcal{B}}$	87	I_P	80
$\overline{f}$	29	ℓ^1	5
$\overline{\mathcal{F}}$	30	ℓ^2	7
$\overline{Q}$	38	ℓ^p	7
$\widehat{Q}$	38	ℓ^∞	5
$\hat{x}$	7	$\mathcal{L}^1$	118
$\hat{X}$	8	$\mathcal{L}^\infty$	118
$(\cdot)^\circ$	35	Λ_∞	31
$\mathcal{A}^c$	108	m_f	16
$\mathcal{A}^{cc}$	108	$\Omega_{\mathcal{B}}$	30
$\wedge B$	26	2^Ω	27
		Px	59
		P_T	90
		ρ_μ	100

Index

Druck: Strauss Offsetdruck, Mörlenbach
Verarbeitung: Schäffer, Grünstadt